MW00396929

Strain Measurements and Stress Analysis

Akhtar S. Khan

Department of Mechanical Engineering
University of Maryland, Baltimore County

Xinwei Wang

Department of Aircraft Engineering
Nanjing University
of Aeronautics and Astronautics

Prentice Hall
Upper Saddle River, New Jersey 07458

Library of Congress Cataloging-in-Publication Data
Khan, Akhtar S.
    Strain measurements and stress analysis / Akhtar S. Khan, Xinwei Wang.
        p.cm.
    Includes bibliographical references and index.
    ISBN 0-13-080076-7
    1. Strain Gages. 2. Strains and stresses—Measurements. 3. Materials—Testing.
    4. Moiré method. I. Wang, Xinwei. II. Title.
    TA413.5.K45 2001
    620.1'123'0287—dc21

                                                    00-042797

Vice President and Editorial Director, ECS: MARCIA HORTON
Acquistions Editor: LAURA CURLESS
Executive Managing Editor: VINCE O'BRIEN
Managing Editor: DAVID A. GEORGE
Production Editor: IRWIN ZUCKER
Manufacturing Manager: TRUDY PISCIOTTI
Manufacturing Buyer: PAT BROWN
Vice President and Director of Production and Manufacturing, ESM: DAVID W. RICCARDI
Director of Creative Services: PAUL BELFANTI
Cover Designer: BRUCE KENSALAAR
Marketing Manager: HOLLY STARK
Marketing Assistant: KAREN MOON

 © 2001 by Prentice Hall
Prentice-Hall, Inc.
Upper Saddle River, New Jersey 07458

Printed in the United States of America

10 9 8 7 6 5 4 3 2 1

ISBN 0-13-080076-7

Prentice-Hall International (UK) Limited, London
Prentice-Hall of Australia Pty. Limited, Sydney
Prentice-Hall Canada Inc., Toronto
Prentice-Hall Hispanoamericana, S.A., Mexico
Prentice-Hall of India Private Limited, New Delhi
Prentice-Hall of Japan, Inc., Tokyo
Pearson Education Asia Pte. Ltd., Singapore
Editora Prentice-Hall do Brasil, Ltda., Rio de Janeiro

Dedicated to Farhan, Meena and Tanveer for their encouragement, support, and patience.

Akhtar S. Khan

谨以此书献给我的家人----妻子张国英，女儿霞萍和儿子熊飞。

Dedicated to my wife Zhang Guoying for her support and my daughter Xiaping and son Xiongfei for their patience.

Xinwei Wang

About the Cover

FRONT COVER
Photoelastic Stress Analysis
(Courtesy of Measurements Group Inc.)

BACK COVER
Stress Analysis of a Bicycle Frame Using Metal-Foil Resistance Rosette
Strain Gages
(Courtesy of Professor Dwayne Arola, UMBC)

Contents

Foreword

With the changing needs of industry for survival under global competitiveness, the traditional science-oriented engineering education for lifelong engineering experiences has become obsolete. Noting that the graduates of U.S. engineering schools lack the needed real-world engineering experiences, the National Science Foundation set out to promote the development of bold, innovative, and comprehensive education reforms in undergraduate engineering education. Among the necessary reform is the synthesis of knowledge for problem solving through hands-on laboratory experiences, which generally requires knowledge in experimental mechanics. This book is the first of its kind to provide not only the fundamentals, but also the practice of strain gage, modern optical, and data acquisition techniques to the upper-division undergraduate students. The abundant solved and unsolved problems throughout the book are related to real problems and should stoke the reader's curiosity, which is essential in the inquiry-based learning process. As noted by the authors, the book can also be used in an introductory graduate-level course, and with references, in an advanced graduate-level course. This book is a welcome addition to the limited selection of modern textbooks on strain measurements and stress analysis.

Albert S. Kobayashi
University of Washington
Seattle, Washington

Preface

This book was developed while teaching a two-credit senior-level elective course at the University of Oklahoma, Norman. Chapters 2, 3, and 4, and part of Chapter 7, are designed for such a course. Additional chapters are included with other courses in mind. For example, a required two-credit course at a junior level at the University of Maryland, Baltimore County will only use Chapters 2 and 3 and part of Chapter 4; an elective senior-level course at the same university will use Chapters 2, 3, 4, 6, and 7. This book can also be used for an introductory graduate-level course, where most of the chapters can be used. It was not the intention of the authors to use this book for an advanced or second-level graduate course; the *Handbook on Experimental Mechanics,* edited by Professor Albert Kobayashi, is an ideal reference book at that level. However, its use is not precluded for such a course, as instructors and students may use the references provided at the end of each chapter as supplements to this book. Chapter 1 is included as a summary of elementary mechanics of materials. The chapter is written in a fairly simple format, with ample solved examples, with the objective that undergraduate students can review the material on their own before starting a laboratory course.

Most two-credit undergraduate courses on experimental mechanics include only one lecture per week and two hours in a laboratory. One hour of classroom instruction per week is normally sufficient *only* to teach the theory behind experimental techniques; it is not sufficient to teach the application of the theory. The requirement of a laboratory report every week somewhat precludes assignments of homework to comprehend application of the theory. Thus it is essential that an undergraduate text should have ample solved examples, worked out in detail, in order for students to learn applications on their own before they enter a laboratory. This should enhance appreciation of their measurements and facilitate the writing of laboratory reports based on these measurements. The need for this book was realized by us when we found no book available in which sufficient solved examples were provided and theory of experimental techniques explained at a level elementary enough for self-reading by undergraduate students if an instructor elected to teach only the applications. This book is intended to fill this vacuum in the field.

Laboratory experiments are not included, as they normally depend on the equipment available at a particular university. However, a supplement to this book is being developed, based on experiments in courses taught at the University of Oklahoma and the University of Maryland, Baltimore County. The supplement should be available soon after this book is printed. The objective of the supplement is to help an instructor who may be designing a new laboratory course or modifying an existing course. This supplement will include about 20 laboratory exercises, from which an instructor can choose those for use in a particular course. A solution manual of the exercises at the end of each chapter is available now.

The authors thank several colleagues who provided valuable comments through their anonymous reviews. We also gratefully acknowledge help from graduate students Haoyue Zhang and Ye Yuan for typing and illustration assistance.

Akhtar S. Khan
Xinwei Wang

About the Authors

Akhtar S. Khan has been Professor of Mechanical Engineering at the University of Maryland, Baltimore County since 1992. Earlier, he taught at the University of Oklahoma for fourteen years and was a Professor of Mechanical Engineering since 1984. He is the founding Editor-in-Chief of the *International Journal of Plasticity,* which is published monthly by Pergamon, a division of Elsevier Science. Dr. Khan was elected a fellow of the American Society of Mechanical Engineers in 1990 and is a member of the Society for Experimental Mechanics and the American Academy of Mechanics. He is the founding chairman of the International Symposium on Plasticity and Its Current Applications, which is now held every 18 months in different parts of the world. Professor Khan has published over 70 papers in leading international journals, mostly in the area of experimental finite plasticity. He is the lead coauthor of *Continuum Theory of Plasticity,* published by John Wiley & Sons, which is widely used at leading universities in the world, Dr. Khan received the Ph.D. degree in mechanics and materials science from the Johns Hopkins University in 1972, after receiving the B.S. degree in mechanical engineering from Aligarh University in India.

Xinwei Wang has been Professor at Nanjing University of Aeronautics and Astronautics in the People's Republic of China since 1992. Dr. Wang earned the B.S. degree in aircraft design and the M.S. degree in solid mechanics in 1975 and 1981, respectively, from the same university. He received the Ph.D. degree in mechanical engineering from the University of Oklahoma in 1989. Professor Wang has been a visiting Associate Professor at the University of Maryland, Baltimore County (1995), a visiting Assistant Professor at the University of Oklahoma (1989–1992), and a Research Associate at UCLA (1984–1985). He is a Director of Chinese Society of Theoretical and Applied Mechanics and a member of the American Society of Mechanical Engineers and the American Academy of Mechanics. He has published over 40 papers in the areas of experimental finite plasticity and finite element techniques.

C H A P T E R   1

Stress, Strain, and Stress–Strain Relationships

1.1 INTRODUCTION

Experimental stress analysts must have a fundamental understanding of stress, strain, and the stress–strain relationships. In this chapter we discuss concepts and equations used most frequently in experimental stress and strain analyses. The book is designed as an elementary text and covers only two-dimensional cases of surface deformation. The measurement of surface deformation is most common; therefore, two-dimensional coverage is sufficient for the experimental stress analyst.

1.2 STRESS

Consider an arbitrary point P on an arbitrary internal or external surface, which may be flat or curved, as illustrated in Fig. 1.1. The stress vector $\mathbf{T}$ at a point P is defined

$$\mathbf{T} = \lim_{\Delta A \to 0} \frac{\Delta \mathbf{F}}{\Delta A} \tag{1.1}$$

where ΔA is the area in the neighborhood of point P on the surface and $\Delta \mathbf{F}$ is the resulting force vector, which may be either a surface force and/or a body force. Note that the stress vector is acting in the direction of the force vector, which does not necessarily coincide with the outward normal $\mathbf{n}$ associated with area ΔA.

It is clear that the stress vector $\mathbf{T}$ depends both on the position P in the body and on the orientation $\mathbf{n}$ of the plane containing the point. When the body is subjected to an arbitrary set of loads, both the magnitude and direction of the stress vector $\mathbf{T}$ at any point vary with the change of orientation of the plane under consideration. The stress

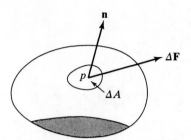

FIGURE 1.1   Resultant force acting on a small surface area.

vector at a point is a meaningless term unless related to a specified plane, since there are an infinite number of planes that may be drawn through any given point.

Since **T** is a vector, it is possible to decompose the stress vector **T** into components called the *stress components*. Consider first that **T** is decomposed into its normal and tangential components, namely,

$$\mathbf{T} = \sigma_n \mathbf{n} + \tau \mathbf{t} \tag{1.2}$$

where **n** and **t** are unit vectors in the normal and tangential directions of the area, as shown in Fig. 1.2. σ_n is the normal stress component, or simply the *normal stress*, and τ is the shear stress component, or simply *shear stress*.

As illustrated in Fig. 1.3, it is also possible to resolve **T** into three Cartesian components for any chosen Cartesian coordinate system. Let the outward normal **n** be parallel to the x axis; then

$$\mathbf{T} = \sigma_{xx} \mathbf{i} + \tau_{xy} \mathbf{j} + \tau_{xz} \mathbf{k} \tag{1.3}$$

where **i, j,** and **k** are the unit vectors along the x axis, y axis, and z axis, respectively, and σ_{xx}, τ_{xy}, and τ_{xz} are the normal stress and shear stresses.

Although there exist an infinite number of stress systems at any point in a body, since there are an infinite number of planes that may be drawn through a point, it is still possible to define the stress state at any point in terms of the stress components acting on the planes of an elemental cube situated at the point. Figure 1.4 shows nine Cartesian stress components acting on the faces of an elemental cube. It is not difficult to show that once these stress components are known, the stress state at that point can be defined completely by equations of stress transformation.

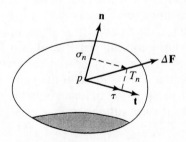

FIGURE 1.2   Resolution of **T** into two components.

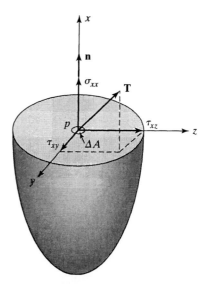

FIGURE 1.3 Resolution of **T** into three Cartesian components.

The subscript and sign conventions employed in Fig. 1.4 are as follows: The first subscript represents the plane where the stress component acts, and the second subscript refers to the direction of the stress component. If the outward normal of the plane is in the direction of the positive x, y, or z axis, the associated positive and negative normal stress components are also in the direction of the positive and negative x, y, or z axis, respectively. Shear stress is assumed positive on a plane with its normal along any positive axis if it is directed along any of the other two positive axes; otherwise, it is negative.

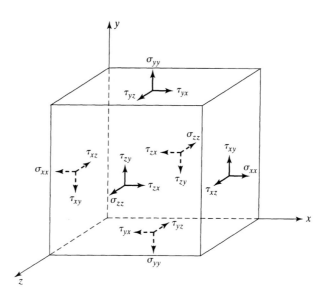

FIGURE 1.4 Cartesian stress components acting on a cube.

1.3   TWO-DIMENSIONAL STRESS STATE AT A POINT

Figure 1.5 shows a free-body diagram for an infinitesimal general two-dimensional case at point P. A unit length in the z direction is assumed throughout this chapter when a two-dimensional free body is considered. Also assume that $\tau_{xz} = \tau_{yz} = \tau_{zx} = \tau_{zy} = 0$ and that σ_{zz} may be zero (e.g., in plane stress cases) or may be different from zero (e.g., in plane strain cases). Only stress components σ_{xx} (or σ_x), σ_{yy} (or σ_y), and τ_{xy} are considered. For simplicity, a single subscript is used for normal stresses since the subscripts are repeated.

The balance of moment about o can be expressed as

$$(\sigma_x dy)\frac{dy}{2} - (\sigma_x dy)\frac{dy}{2} + (\sigma_y dx)\frac{dx}{2} - (\sigma_y dx)\frac{dx}{2} \tag{1.4}$$
$$- (\tau_{xy} dy)\,dx + (\tau_{yx} dx)\,dy = 0$$

Since $dx\,dy$ is not zero, Eq. (1.4) reduces to

$$\tau_{xy} = \tau_{yx} \tag{1.5}$$

The shear stress component acting on the x plane in the y direction is equal to the shear stress component acting on the y plane in the x direction. This result holds for any two perpendicular planes (say x' plane and y' plane). Thus $\tau_{x'y'} = \tau_{y'x'}$, where $\tau_{x'y'}$ and $\tau_{y'x'}$ are shear stresses acting on the x' plane in the y' direction and y' plane in the x' direction, respectively.

Consider next the force equilibrium in the x and y directions on the free body shown in Fig. 1.6a. If the inclined surface equals A, then

$$\Sigma F_x = 0$$
$$(\sigma_x' A)\cos\alpha - (\tau_{x'y'} A)\sin\alpha - (\sigma_x A\cos\alpha) - (\tau_{yx} A\sin\alpha) = 0$$
$$\Sigma F_y = 0 \tag{1.6}$$
$$(\sigma_x' A)\sin\alpha + (\tau_{x'y'} A)\cos\alpha - (\tau_{xy} A\cos\alpha) - (\sigma_y A\sin\alpha) = 0$$

By solving Eq. (1.6) and noting that $\tau_{xy} = \tau_{yx}$, we find that

$$\sigma_{x'} = \sigma_x \cos^2\alpha + \sigma_y \sin^2\alpha + 2\tau_{xy}\sin\alpha\cos\alpha \tag{1.7}$$

$$\tau_{x'y'} = (\sigma_y - \sigma_x)\sin\alpha\cos\alpha + \tau_{xy}(\cos^2\alpha - \sin^2\alpha) \tag{1.8}$$

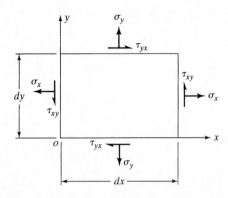

FIGURE 1.5   Stress components in two dimensions.

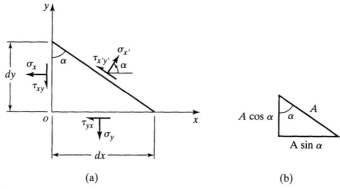

FIGURE 1.6   Free body showing the stress components on an inclined plane.

The stress $\sigma_{y'}$ may be found simply by substituting $(\alpha + \pi/2)$ for α in Eq. (1.7). Note that

$$\sin\left(\alpha + \frac{\pi}{2}\right) = \cos\alpha$$

and

$$\cos\left(\alpha + \frac{\pi}{2}\right) = -\sin\alpha$$

so we obtain

$$\sigma_y' = \sigma_y\cos^2\alpha + \sigma_x\sin^2\alpha - 2\tau_{xy}\sin\alpha\cos\alpha \qquad (1.9)$$

Equations (1.7) to (1.9) are known as the *equations of stress transformation.* They can be employed to determine the stress components $\sigma_{x'}$, $\sigma_{y'}$, and $\tau_{x'y'}$ associated with any Cartesian coordinate system $x'y'$ once σ_x, σ_y, τ_{xy}, and α are known. For any two-dimensional case, the stress state at a point is completely determined if the stress components on any two perpendicular planes passing through the point are known, since the stress components on any other planes that pass through the point can be found by the equations of stress transformation once α is specified.

EXAMPLE 1.1

The Cartesian stress components at a point are shown in Fig. E1.1. Determine the normal and shear stresses on the plane indicated.

Solution

$$\tan\alpha_1 = \frac{5}{12} \qquad \alpha_1 = 22.62°$$

Thus the angle α between the outward normal of the plane and the x axis is given by

$$\alpha = 90° + \alpha_1 = 112.62°$$

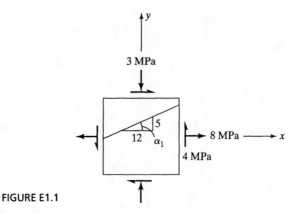

FIGURE E1.1

By using Eq. (1.7) and noting that $\sigma_x = 8$ MPa, $\sigma_y = -3$ MPa, and $\tau_{xy} = 4$ MPa, one obtains

$$\sigma = 8 \cos^2 112.62° + (-3)\sin^2 112.62° + (2)(4) \sin 112.62° \cos 112.62°$$
$$= -4.21 \text{ MPa} \quad (\text{compression})$$

Similarly, Eq. (1.8) gives

$$\tau = [(-3)-8] \sin 112.62° \cos 112.62° + 4(\cos^2 112.62° - \sin^2 112.62°)$$
$$= 1.09 \text{ MPa}$$

Adding Eqs. (1.7) and (1.9) yields

$$\sigma_{x'} + \sigma_{y'} = \sigma_x + \sigma_y \tag{1.10}$$

Therefore, the sum of the normal stresses on any two perpendicular planes is an invariant quantity.

EXAMPLE 1.2

The stress components at a point are shown in Fig. E1.2. If $\sigma_x' = 3$ ksi, find α and $\tau_{x'y'}$.

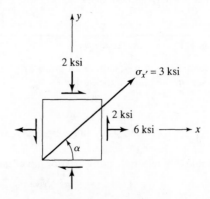

FIGURE E1.2

Solution
Equation (1.10) gives

$$\sigma_{y'} = \sigma_x + \sigma_y - \sigma_{x'} = 6 + (-2) - 3 = 1 \text{ ksi}$$

Using Eqs. (1.7) and (1.9), we obtain

$$\sigma_{x'} - \sigma_{y'} = (\sigma_x - \sigma_y) \cos 2\alpha + 2\tau_{xy} \sin 2\alpha$$
$$3 - 1 = (6 + 2) \cos 2\alpha + (2)(2) \sin 2\alpha$$

or

$$1 - 2 \sin 2\alpha = 4(1 - \sin^2 2\alpha)^{1/2}$$
$$1 - 4 \sin 2\alpha + 4 \sin^2 2\alpha = 16 - 16 \sin^2 2\alpha$$

or

$$20 \sin^2 2\alpha - 4 \sin 2\alpha - 15 = 0 \qquad\qquad (\text{E1.2a})$$

Solving Eq. (E1.2a) yields

$$\sin 2\alpha = \frac{4 \pm [16 + 4(15)(20)]^{1/2}}{40} = \frac{1 \pm (1 + 75)^{1/2}}{10}$$

when $\sin 2\alpha = -0.7718$, $\alpha_1 = -25.26°$ (CW) or $\alpha_1 = -64.74°$ (CW); when $\sin 2\alpha = 0.9718$, $\alpha_2 = 38.18°$ (CCW) or $\alpha_2 = 51.82°$ (CCW), where CW and CCW mean clockwise and counterclockwise, respectively. It can be shown that $-64.74°$ and $38.18°$ are not correct answers, so they are discarded.

For $\alpha_1 = -25.26°$, Eq. (1.8) gives

$$\tau_{x'y'} = (-2 - 6) \sin(-25.26°) \cos(-25.26°) + 2[\cos^2(-25.26°) - \sin^2(25.26°)] = 4.36 \text{ ksi}$$

For $\alpha_2 = 51.82°$, Eq. (1.8) gives

$$\tau_{x'y'} = (-2 - 6) \sin 51.82° \cos 51.82° + 2(\cos^2 51.82° - \sin^2 51.82°)$$
$$= -4.36 \text{ ksi}$$

1.4 PRINCIPAL STRESSES

To determine the planes on which the maximum and minimum normal stresses act, it is convenient to rewrite Eqs. (1.7) to (1.9) in terms of 2α. Using the trigonometric identities

$$\sin 2\alpha = 2 \sin \alpha \cos \alpha \qquad \sin^2\alpha = \frac{1 - \cos 2\alpha}{2} \qquad \cos^2\alpha = \frac{1 + \cos 2\alpha}{2}$$

Eqs. (1.7) to (1.9) become

$$\sigma_x' = \frac{\sigma_x + \sigma_y}{2} + \frac{\sigma_x - \sigma_y}{2} \cos 2\alpha + \tau_{xy} \sin 2\alpha \qquad\qquad (1.7a)$$

$$\tau_{x'y'} = \frac{\sigma_y - \sigma_x}{2} \sin 2\alpha + \tau_{xy} \cos 2\alpha \qquad\qquad (1.8a)$$

$$\sigma_y' = \frac{\sigma_x + \sigma_y}{2} - \frac{\sigma_x - \sigma_y}{2} \cos 2\alpha - \tau_{xy} \sin 2\alpha \qquad\qquad (1.9a)$$

Differentiating Eq. (1.7a) with respect to α and setting the derivative equal to zero, the orientation of the planes on which maximum and minimum normal stresses act can be computed as

$$\tan 2\alpha = \frac{2\tau_{xy}}{\sigma_x - \sigma_y} \tag{1.11a}$$

Equation (1.11a) has two roots, which are 90° apart. A comparison of Eqs. (1.8a) and (1.11a) indicates that the shear stress on these planes is zero. The planes on which the shear stress vanishes are called the *principal planes;* the normal stresses on the principal planes are called the *principal stresses.*

The principal stresses can be calculated using the following equations:

$$\sigma_1 = \sigma_{\max} = \frac{\sigma_x + \sigma_y}{2} + \left[\left(\frac{\sigma_x - \sigma_y}{2} \right)^2 + \tau_{xy}^2 \right]^{1/2}$$

$$\sigma_2 = \sigma_{\min} = \frac{\sigma_x + \sigma_y}{2} - \left[\left(\frac{\sigma_x - \sigma_y}{2} \right)^2 + \tau_{xy}^2 \right]^{1/2} \tag{1.12}$$

Adding Eqs. (1.12) gives

$$\sigma_1 + \sigma_2 = \sigma_x + \sigma_y$$

which again shows that the sum of normal stresses on any two perpendicular planes is an invariant.

It is not difficult to show that the particular value of α in Eq. (1.11a) can be determined as

$$
\begin{aligned}
0 &< \alpha < \pi/2 & & \text{if } \tau_{xy} > 0 \\
\frac{-\pi}{2} &< \alpha < 0 \quad \text{or} \quad \frac{\pi}{2} < \alpha < \pi & & \text{if } \tau_{xy} < 0
\end{aligned} \tag{1.13}
$$

where α is the angle between the x axis and the direction of σ_1 and is called the *principal angle.* For convenience, α is considered positive if the direction of the σ_1 axis is counterclockwise from the x axis. In practice, it is more convenient to compute the principal angles by the following equation (see reference [1]) than by Eqs. (1.11a) and (1.13):

$$\alpha_i = \tan^{-1} \frac{\sigma_i - \sigma_x}{\tau_{xy}} = \tan^{-1} \frac{\tau_{xy}}{\sigma_i - \sigma_y}$$

where α_i, σ_i $(i = 1,2)$, σ_x, σ_y, and τ_{xy} are the principal angle, the principal stress, the normal stress in the x direction, the normal stress in the y direction, and the shear stress, respectively.

EXAMPLE 1.3

A thin-walled pressure vessel is subjected to an internal pressure of 200 psi and a torque of 20,000 in.-lb, as shown in Fig. E1.3a. The thickness and mean radius of the vessel are 0.1 in. and 4 in. Determine the principal stresses and principal angle with respect to the x axis (the axial axis) at point A.

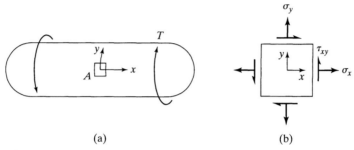

(a)   (b)

FIGURE E1.3

Solution

$$\sigma_x = \frac{PR}{2t} = \frac{(200)(4)}{(2)(0.1)} = 4 \text{ ksi}$$

$$\sigma_y = \frac{PR}{t} = 2\sigma_x = 8 \text{ ksi}$$

$$\tau_{xy} = \frac{T}{2\pi R^2 t} = \frac{20,000}{2\pi(4^2)(0.1)} = 1.99 \text{ ksi}$$

Thus

$$\frac{\sigma_x + \sigma_y}{2} = \frac{4 + 8}{2} = 6 \text{ ksi}$$

$$\left[\frac{(\sigma_x - \sigma_y)^2}{4} + \tau_{xy}^2\right]^{1/2} = \left[\frac{(4-8)^2}{4} + 1.99^2\right]^{1/2} = 2.82 \text{ ksi}$$

Using Eqs. (1.11a) and (1.12), we obtain

$$\sigma_1 = 6 + 2.82 = 8.82 \text{ ksi}$$
$$\sigma_2 = 6 - 2.82 = 3.18 \text{ ksi}$$
$$\alpha = \frac{1}{2}\tan^{-1}\frac{2\tau_{xy}}{\sigma_x - \sigma_y} = 0.5 \tan^{-1}(-0.995) = -22.4°$$

Since $\tau_{xy} > 0$ and $0 < \alpha_1 < \pi/2$,

$$\alpha_1 = 90° - 22.4° = 67.6° \quad \text{(CCW)}$$

Or simply, using Eq. (1.11b),

$$\alpha_1 = \tan^{-1}\frac{\tau_{xy}}{\sigma_1 - \sigma_y} = \tan^{-1}(2.427) = 67.6° \quad \text{(CCW)}$$

If one differentiates Eq.(1.8a) with respect to α and sets the derivative equal to zero, the orientation of the planes on which maximum shear stress acts can be obtained as

$$\tan 2\alpha = \frac{-(\sigma_x - \sigma_y)}{2\tau_{xy}} \tag{1.14}$$

The two roots of α in Eq. (1.14) also define a set of perpendicular planes on which the shear stress is maximum. By comparing Eq. (1.14) with Eq. (1.11a), we find that they are negative reciprocals. Therefore, the roots of 2α from these two equations are 90° apart, or the corresponding planes are 45° apart, as shown in Fig. 1.7.

From Eqs. (1.8a), (1.14), and (1.12), the maximum shear stress can be determined

$$\tau_{max} = \left[\left(\frac{\sigma_x - \sigma_y}{2} \right)^2 + \tau_{xy}^2 \right]^{1/2} = \frac{\sigma_1 - \sigma_2}{2} \tag{1.15a}$$

It can also be shown that the normal stresses acting on these planes are equal and given by

$$\sigma_m = \frac{\sigma_x + \sigma_y}{2} \tag{1.15b}$$

Several stress loci in a two-dimensional stress field are described below. These are useful in experimental stress analysis.

Isostatic or Principal-Stress Trajectory. *Isostatic* is a term that describes a line where at any point a maximum or minimum principal stress axis is tangent to the line. Since the two principal stresses at any point are mutually perpendicular, the isostatics form two orthogonal families of curves, one that represents the maximum principal stress trajectory and the other one represents the minimum principal stress trajectory.

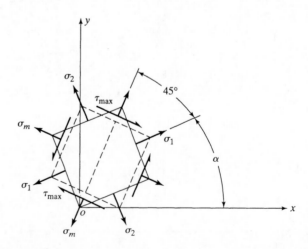

FIGURE 1.7   Planes of maximum shear stress and principal stresses.

Isoclinic. *Isoclinic* is a term that describes a line along which the direction of the principal stresses is constant. It is obvious that the straight isostatic must also be the isoclinic.

Isochromatic. *Isochromatic* is a term that describes a line along which the difference between the maximum and minimum principal stresses, $\sigma_1 - \sigma_2$, is a constant. The terms *isoclinic* and *isochromatic* will both be encountered in discussions of photoelasticity. The term *isochromatic* perhaps derives from the fact that it represents a constant color band in photoelasticity.

Isopachic. *Isopachic* is a term that describes a line along which the sum of the maximum and minimum principal stresses, $\sigma_1 + \sigma_2$, is a constant. The value may be determined by membrane, shadow moiré, or electric field analog methods.

There are several other stress loci, such as isobar and isoentatic. These and other loci are discussed in detail in reference [2].

1.5 MOHR'S STRESS CIRCLE

The equations of stress transformation can conveniently be represented by a circle, called *Mohr's stress circle,* developed by O. Mohr (1835–1918), an outstanding German professor of structural mechanics. Mohr's stress circle provides a powerful aid to visualization of a stress state at a point.

Rewriting Eq. (1.7a), we obtain

$$\sigma_{x'} - \frac{\sigma_x + \sigma_y}{2} = \frac{\sigma_x - \sigma_y}{2}\cos 2\alpha + \tau_{xy}\sin 2\alpha \tag{1.7b}$$

Squaring Eqs. (1.7b) and (1.8a) and adding them together yields

$$\left(\sigma - \frac{\sigma_x + \sigma_y}{2}\right)^2 + \tau^2 = \left(\frac{\sigma_x - \sigma_y}{2}\right)^2 + \tau_{xy}^2 \tag{1.16}$$

where $\sigma_{x'}$ has been replaced by σ and $\tau_{x'y'}$ has been replaced by τ, respectively. Obviously, Eq. (1.16) is the equation of a circle in the σ–τ coordinate system with its center at $((\sigma_x + \sigma_y)/2, 0)$ and a radius of $[(\sigma_x - \sigma_y)^2/4 + \tau_{xy}^2]^{1/2}$. This is Mohr's stress circle, an example of which is shown later in Fig. E1.4b.

The sign conventions in drawing Mohr's stress circle are selected as follows: The normal stresses are considered positive if they are tensile and are usually plotted to the right side of the origin; the shear stress is considered positive if it produces a clockwise moment about the center of the element and is plotted above the horizontal axis (the σ axis). For example, the shear stress τ_{xy} on the x face in Fig. E1.4a produces a counterclockwise moment, so it is considered negative; on the other hand, the shear stress τ_{yx} on the y face in Fig. E1.4a produces a clockwise moment, so it is considered positive. In short, point x is at $(\sigma_x, -\tau_{xy})$ and point y is at (σ_y, τ_{yx}) if the stress components shown in Fig. 1.5 are considered positive.

In general, every point on Mohr's stress circle represents a plane that passes through the position at which the stress state is known. The angle between the radii drawn to two points on the circle is twice the angle in space between the corresponding planes, which is measured in the same direction as the double angle on Mohr's stress circle if the sign conventions described above are adopted. For example, if point 1 is counterclockwise from x by 2α as shown in Fig. E1.4b, the direction of the maximum principal plane is also counterclockwise from x by α, as shown in Fig. E1.4c.

Example 1.4 illustrates the general procedure to follow in constructing a Mohr's stress circle to find the principal stresses, principal angle, and maximum shear stress.

EXAMPLE 1.4

Consider the stresses at a critical point in the engineering structural member depicted in Fig. E1.4a. Sketch Mohr's stress circle and find the principal stresses, principal angle, and maximum shear stress.

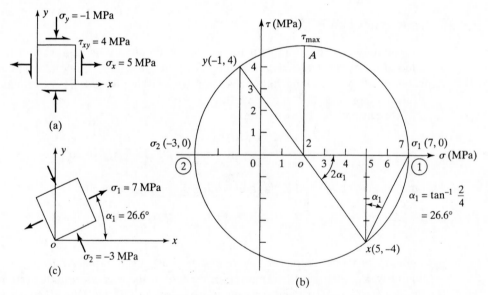

FIGURE E1.4 Mohr's stress circle.

Solution

Step 1: Set up the σ–τ coordinate system where one division is chosen to represent 1 MPa for both σ and τ axes. Note that the scale should always be the same for both axes. Use of engineering paper is recommended.

Step 2. Locate the points that represent the stresses on plane x and plane y of the element. In accordance with standard sign conventions, point x is at $\sigma = \sigma_x = +5$ MPa and $\tau = \tau_{xy} = -4$ MPa (counterclockwise moment) and point y is at $\sigma = \sigma_y = -1$ MPa and $\tau = \tau_{yx} = +4$ MPa (clockwise moment), as shown in Fig. E1.4b.

Step 3. Connect points x and y with a line that intersects the σ axis at point o. Draw a circle with the center at o and with a radius of ox (or oy), as shown in Fig. E1.4b. This is Mohr's stress circle.

Step 4. Analysis of Mohr's circle
(a) Points 1 and 2 are principal planes. The stress at point 1 is the maximum principal stress σ_1. From Fig. E1.4b it can be found that σ_1 is approximately equal to 7 MPa (7 divisions). Similarly, σ_2 is approximately equal to −3 MPa (3 divisions but on the negative side). Note that only approximate values can be obtained by this graphical method.
(b) The angle between ox and $o1$ equals approximately 53°, which is twice the principal angle. The principal angle α therefore equals +26.5°(counterclockwise direction). The principal stress state is shown in Fig. E1.4c.
(c) Point A represents the plane on which the maximum shear stress acts. Note that it is 90° from point 1. Remember that α on the element corresponds to 2α on the circle. It is clear that the plane on which the maximum shear stress acts is always 45° (counterclockwise direction) from the principal plane where σ_1 acts. The magnitude of the maximum shear stress is equal to the radius, which is approximately 5 MPa (5 divisions). In summary, $\sigma_1 = +7$ MPa, $\sigma_2 = -3$ MPa, $\alpha = +26.5°$, and $\tau_{max} = +5$ MPa. Note that Eqs. (1.11) to (1.13) and (1.15) should be used when accurate values are required, as illustrated in Example 1.3.

1.6 DIFFERENTIAL EQUATIONS OF EQUILIBRIUM

Thus far the stress state at a point, or rather the stress state on a small cube surrounding the point, has been considered. The cube was assumed to be so small that the stresses did not vary significantly from one face to the next. Let us now consider the equilibrium of a rectangular solid with side lengths of δx and δy, paralleling the x and y axes, whose unit length is in the z direction. The side lengths of δx and δy are small, but stresses can change significantly from face to face, as shown in Fig. 1.8. This case contrasts with the one considered previously.

In Fig. 1.8, X and Y are components of the body force intensity in the x and y directions, which are assumed to vary insignificantly within the small rectangular solid

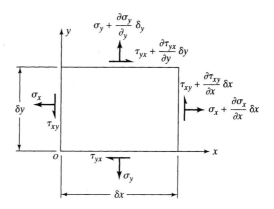

FIGURE 1.8 Stress distributions on an element.

under consideration. First equating the sum of all forces to zero for static equilibrium in the x direction, one obtains

$$X\delta x\,\delta y + \left(\sigma_x + \frac{\partial\sigma_x}{\partial x}\delta X\right)\delta y - \sigma_x\,\delta_y + \left(\tau_{yx} + \frac{\partial\tau_{yx}}{\partial y}\delta y\right)\delta x - \tau_{yx}\,\delta x = 0 \qquad (1.17)$$

which can be simplified as

$$\left(\frac{\partial\sigma_x}{\partial x} + \frac{\partial\tau_{yx}}{\partial y} + X\right)\delta x\,\delta y = 0 \qquad (1.18)$$

Since $\delta x\,\delta y$ is not zero, the sum in parentheses must vanish, namely,

$$\frac{\partial\sigma_x}{\partial x} + \frac{\partial\tau_{yx}}{\partial y} + X = 0 \qquad (1.19)$$

Similarly, equating the sum of all forces in the y direction to zero yields

$$\frac{\partial\tau_{xy}}{\partial x} + \frac{\partial\sigma_y}{\partial y} + Y = 0 \qquad (1.20)$$

In many problems, the body forces are unimportant and can be ignored. Using Eq. (1.5), Eqs. (1.19) and (1.20) can be simplified further:

$$\begin{aligned}\frac{\partial\sigma_x}{\partial x} + \frac{\partial\tau_{xy}}{\partial y} &= 0 \\[2mm] \frac{\partial\tau_{xy}}{\partial x} + \frac{\partial\sigma_y}{\partial y} &= 0\end{aligned} \qquad (1.21)$$

The *Lamé–Maxwell equations* represent another useful way to express equilibrium conditions in two dimensions. They are expressed in terms of principal stresses in a curvilinear coordinate system formed by the principal stress trajectories s_1 and s_2. Consider the equilibrium of a small element with unit thickness bounded by adjacent stress trajectories, as shown in Fig. 1.9. According to the definition of principal stress trajectory, the boundaries of the element will be free of shear stress, so no shear stress terms will appear in the equations.

Consider first the side lengths of the element. Let $AB = \delta s_2$; then to a first degree of approximation, the length of CD is equal to $\delta s_2 + \delta s_1\,\delta\theta_2$. Similarly, if the length of AD is δs_1, the length of BC is approximately equal to $\delta s_1 - \delta s_2\,\delta\theta_1$. The radii of curvature of sides BC and AB in Fig. 1.9a can be expressed as

$$\frac{\delta S_1}{\delta\theta_1} = \rho_1 \qquad \frac{\delta S_2}{\delta\theta_2} = \rho_2 \qquad (1.22)$$

The positive curvature is defined as in Fig. 1.9b. Equating to zero the sum of all forces parallel to the x axis yields

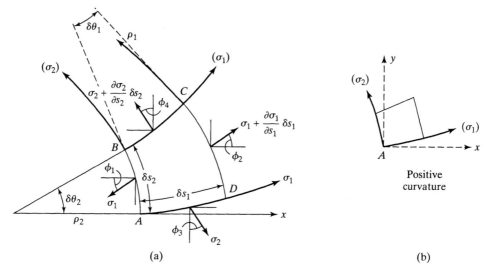

FIGURE 1.9   Stress distributions on an element.

$$\left(\sigma_1 + \frac{\partial \sigma_1}{\partial S_1}\,\delta S_1\right)(\delta S_2 + \delta S_1\,\delta\theta_2)\cos\phi_2 - \sigma_1\,\delta S_2\cos\phi_1 + \sigma_2\,\delta S_1\sin\phi_3$$

$$-\left(\sigma_2 + \frac{\partial \sigma_2}{\partial S_2}\,\delta S_2\right)(\delta S_1 - \delta S_2\,\delta\theta_1)\sin\phi_4 = 0 \tag{1.23}$$

If the element is small enough, the value of all angles designated ϕ_i should be small and $\sin\phi_i \approx \phi_i$ and $\cos\phi_i \approx 1$, since the x and y axes are tangential to the σ_1 and σ_2 trajectories at the element corner, as shown in Fig. 1.9b. By neglecting third- and fourth-order terms, Eq. (1.23) can be simplified as

$$\sigma_1\,\delta S_1\,\delta\theta_2 + \frac{\partial \sigma_1}{\partial S_1}\,\delta S_1\,\delta S_2 + \sigma_2\,\delta S_1(\phi_3 - \phi_4) = 0 \tag{1.24}$$

Since $\phi_4 - \phi_3$ represents the rotation of the tangent to the σ_2 trajectory from AD to BC, $\phi_4 - \phi_3 \approx \delta\theta_2$. Substituting this equation in Eq. (1.24), using Eq. (1.22), and noting that $\delta s_1\,\delta s_2$ is not zero, one obtains

$$\frac{\partial \sigma_1}{\partial S_1} + \frac{\sigma_1 - \sigma_2}{\rho_2} = 0 \tag{1.25}$$

Similarly, equating the sum of all forces in the y direction to zero, one can finally obtain

$$\frac{\partial \sigma_2}{\partial S_2} + \frac{\sigma_1 - \sigma_2}{\rho_1} = 0 \tag{1.26}$$

Note that if $\rho_2 = \infty$ or $\sigma_1 = \sigma_2$, then σ_1 is at a maximum or minimum, and if $\rho_1 = \infty$ or $\sigma_1 = \sigma_2$, then σ_2 is at a maximum or minimum.

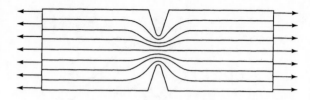

FIGURE 1.10 Isostatics in a V-shaped notched plate.

Equations (1.25) and (1.26), known as the Lamé–Maxwell equations, are also useful in photoelasticity. Lamé–Maxwell equations of equilibrium illustrate that principal stress magnitudes are intimately connected with the shape of stress trajectories in a loaded member. Therefore, any discontinuity of shape in a structure must be associated with rapid changes in stress magnitude called *stress concentrations*, because the discontinuity distorts the stress trajectories. An example is shown in Fig. 1.10. Stress concentrations play an important role in designing structural members. Numerous analytical, numerical, and experimental investigations have been done and the *stress concentration factors* have been extracted and published in either graphical or equation format [1,5]. The stress concentration factor K is defined as

$$K = \frac{\sigma_{max}}{\sigma_0} \tag{1.27}$$

where σ_{max} is the peak or maximum stress (refer to Fig. 1.11a) and σ_0 is the average stress, respectively.

The average stress is usually determined by using the net area of the cross section. For example, in Fig. 1.11a,

$$\sigma_0 = \frac{P}{A_{net}} = \frac{P}{(h - 2r)d} \tag{1.28a}$$

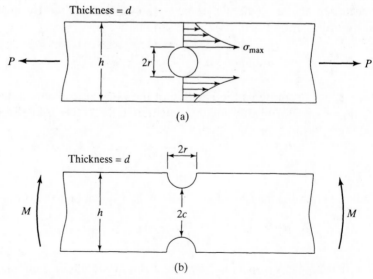

(a)

(b)

FIGURE 1.11   (a) Circular hole plate; (b) semicircular groove plate.

In Fig. 1.11b,

$$\sigma_0 = \frac{MC}{I_{net}} = \frac{MC}{[d(2c)^3]/12} = 1.5\frac{M}{dc^2} \tag{1.28b}$$

Once σ_0 and K are known, σ_{max} can be calculated by using Eq. (1.27). It should be emphasized that the definition of σ_0 may differ from those given in Eq. (1.28); sometimes the average stress is based on the cross-sectional area away from the discontinuity under consideration.

1.7   STRAIN AND DISPLACEMENT

First consider the axial strain for uniaxial cases. It is common knowledge that a bar with an initial length of l_0 will be elongated to l under the tensile load P, as shown in Fig. 1.12. The *engineering, nominal,* or *Lagrangian strain* is defined as the deformation per unit initial length and can be written as

$$\epsilon = \frac{l - l_0}{l_0} = \frac{l}{l_0} - l$$

Obviously, the strain ϵ is positive if the load produces an increase in length (i.e., the positive strain is associated with a tensile stress).

There are several other definitions of strain. For example, *true, logarithmic,* or *natural strain* is defined in the uniaxial case as

$$e = \int_{l_0}^{l} \frac{dl}{l} = \ln\frac{l}{l_0}$$

The difference between these definitions is negligible in the small-strain region. However, when large deformations are involved, the difference becomes appreciable.

Consider now the deformation in a two-dimensional case. Figure 1.13 shows an infinitesimal rectangular element with sides of dx and dy subjected to an infinitesimal deformation. If u and v are the x and y components of displacement at point A, the components of displacement at point B will be $u + (\partial u/\partial x)\,dx$ and $v + (\partial v/\partial x)\,dx$, and the components of displacement at point D will be $u + (\partial u/\partial y)\,dy$ and $v + (\partial v/\partial y)\,dy$, respectively. As in the uniaxial case, the engineering strains in the x and y directions are defined as follows:

$$\begin{aligned}
\epsilon_x &= \frac{A'B' - AB}{AB} = \frac{A'B' - dx}{dx} \\
\epsilon_y &= \frac{A'D' - AD}{AD} = \frac{A'D' - dy}{dy}
\end{aligned} \tag{1.29}$$

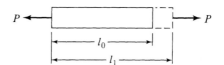

FIGURE 1.12   Deformation in a tensile bar.

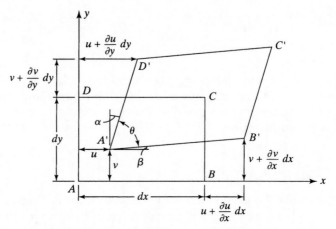

FIGURE 1.13 Deformation of a rectangular element.

The *engineering shear strain* is defined as the change in angle of the initial right angle, namely,

$$\gamma_{xy} = 90° - \theta = \alpha + \beta \tag{1.30}$$

where θ, α, and β are as shown in Fig. 1.13. It is clear that the shear strain is positive when θ is positive but less than 90°. For small deformations, α and β are usually very small, so that $\tan\alpha \approx \alpha$ and $\tan\beta \approx \beta$. Also, $A'B' \approx dx + (\partial u/\partial x)\, dx$ and $A'D' \approx dy + (\partial v/\partial y)\, dy$; therefore, Eqs. (1.29) and (1.30) can be rewritten as

$$\epsilon_x = \frac{\partial u}{\partial x}$$

$$\epsilon_y = \frac{\partial v}{\partial y} \tag{1.31}$$

$$\gamma_{xy} = \frac{\partial u}{\partial y} + \frac{\partial v}{\partial x}$$

which represent the strain–displacement relationships in two dimensions.

It is observed that from a given displacement field u and v, a unique strain field can be determined by using Eq. (1.31). However, for an arbitrary given strain field, an unrealistic displacement field might be obtained (e.g., the body might have voids after deformation). To avoid this and to ensure a valid displacement field, the compatibility equation must be used. For two dimensions, only one compatibility equation is required. That is,

$$\frac{\partial^2 \gamma_{xy}}{\partial x\,\partial y} = \frac{\partial^2 \epsilon_y}{\partial x^2} + \frac{\partial^2 \epsilon_x}{\partial y^2} \tag{1.32}$$

The average rigid-body rotation of the element is defined as follows:

$$\omega = \frac{1}{2}\left(\frac{\partial v}{\partial x} - \frac{\partial u}{\partial y}\right) \tag{1.33}$$

This is useful when measurements are done by the moiré method.

1.8   PRINCIPAL STRAINS AND MOHR'S STRAIN CIRCLE

As shown earlier, in two dimensions the stress state at a point is determined uniquely if the stress components on two planes are known. The same is true for the strain state at a point. Many engineering problems require determining the strain state associated with a particular coordinate system. Therefore, it is necessary to find the relationships between the strains associated with one coordinate system and the strains associated with another coordinate system, or to derive the equations of strain transformation. All derivations in this discussion are restricted to two dimensions, since the most common measurements are for surface deformation.

Suppose that strains ϵ_x, ϵ_y, and γ_{xy} are known and one wishes to find the strains associated with the x'–y' coordinate system, as shown in Fig. 1.14a. The x'–y' coordinate system makes an angle α with the x–y coordinate system. Consider first the strain $\epsilon_{x'}$ in the x' direction, which is

$$\epsilon_{x'} = \frac{ac' - ac}{ac} = \frac{c'c''}{ds}$$

From Fig. 1.14a one finds that $c'c''$ is the component sum of all deformations ($\epsilon_x\, dx$, $\epsilon_y\, dy$, and $\gamma_{xy}\, dy$) in the x' direction. The components in the x' direction are represented by the sides (small arrows) of triangles 1, 2, and 3 parallel to the x' axis, which are $\epsilon_x\, dx$ $\cos \alpha$, $\epsilon_y\, dy \sin \alpha$, and $\gamma_{xy}(1 + \epsilon_y)\, dy \cos \alpha$ ($\approx \gamma_{xy}\, dy \cos \alpha$), respectively, since $\gamma_{xy}\epsilon_y$ can be neglected for small deformations. It is considered positive when the arrow points to the positive x' direction. Therefore,

$$\epsilon_{x'} = \frac{\epsilon_x\, dx \cos \alpha + \epsilon_y\, dy \sin \alpha + \gamma_{xy}\, dy \cos \alpha}{ds}$$

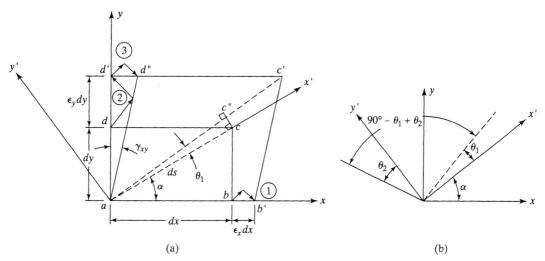

(a)   (b)

FIGURE 1.14   Deformations of a rectangular element in two dimensions.

By identifying that $\sin \alpha = dy/ds$ and $\cos \alpha = dx/ds$, one finally obtains

$$\epsilon_{x'} = \epsilon_x \cos^2\alpha + \epsilon_y \sin^2\alpha + \gamma_{xy} \sin \alpha \cos \alpha \qquad (1.34)$$

The normal strain in the y' direction is found simply by substituting $(90° + \alpha)$ for α in Eq. (1.34). That yields

$$\epsilon_{y'} = \epsilon_x \sin^2\alpha + \epsilon_y \cos^2\alpha - \gamma_{xy} \sin \alpha \cos \alpha \qquad (1.35)$$

By definition, the shear strain in the x'–y' coordinate system is equal to $\theta_1 - \theta_2$, where θ_1 is the small angular change of the line initially parallel to the x' axis and θ_2 is the small angular change of the line initially parallel to the y' axis, as shown in Fig. 1.14b. The angle θ_1 (or θ_2) is very small since the deformations considered here are assumed infinitesimal. Thus

$$\theta_1 = \frac{cc''}{ac} = \frac{cc''}{ds}$$

As in the case of finding the normal strain, cc'' is the component sum of all deformations ($\epsilon_x\,dx$, $\epsilon_y\,dy$, and $\gamma_{xy}\,dy$) in the y' direction. The components in the y' direction are represented by the sides (small arrows) of triangles 1, 2, and 3 parallel to the y' axis, which are $-\epsilon_x\,dx \sin \alpha$, $\epsilon_y\,dy \cos \alpha$, and $-\gamma_{xy}\,dy \cos \alpha$, respectively. It is considered positive when the arrow points to the y' positive direction, as illustrated in Fig. 1.14a. Thus

$$\theta_1 = \frac{-\epsilon_x\,dx \sin \alpha + \epsilon_y\,dy \cos \alpha - \gamma_{xy}\,dy \sin \alpha}{ds}$$

$$= -\epsilon_x \sin \alpha \cos \alpha + \epsilon_y \sin \alpha \cos \alpha - \gamma_{xy} \sin^2\alpha$$

EXAMPLE 1.5

Using equations of strain transformation, determine the normal strain ϵ in direction n, shown in Fig. E1.5, for the given state of strain, namely, $\epsilon_x = 4000\ \mu\epsilon$, $\epsilon_y = -500\mu\epsilon$, and $\gamma_{xy} = 1000\mu\epsilon$.

Solution
Since $\alpha = 35°$, by Eq. (1.34), we have

$$\epsilon_n = (4000) \cos^2 35° + (-500) \sin^2 35° + (1000) \sin 35° \cos 35°$$
$$= 2989.4\mu\epsilon$$

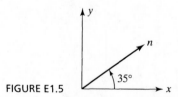

FIGURE E1.5

To determine θ_2, we observe that θ_2 will be given by the equation above if α is replaced by $(90° + \alpha)$, since the y' axis was originally at an angle of $90° + \alpha$ from the x axis. Therefore,

$$\theta_2 = \epsilon_x \sin \alpha \cos \alpha - \epsilon_y \sin \alpha \cos \alpha - \gamma_{xy} \cos^2\alpha$$

The shear strain $\gamma_{x'y'}$ can be found as follows:

$$\gamma_{x'y'} = \theta_1 - \theta_2 = 2(\epsilon_y - \epsilon_x) \sin \alpha \cos \alpha + \gamma_{xy}(\cos^2\alpha - \sin^2\alpha) \qquad (1.36)$$

Equations (1.34) to (1.36) are called the *equations of strain transformation*. Note that the angle α is positive when measured counterclockwise (CCW) from the x axis. Adding Eqs. (1.34) and (1.35) gives

$$\epsilon_{x'} + \epsilon_{y'} = \epsilon_x + \epsilon_y$$

It can be shown that the algebraic sum of the normal strains at a point in any two mutually perpendicular directions is an invariant, namely,

$$\epsilon_{x'} + \epsilon_{y'} = \epsilon_x + \epsilon_y = \epsilon_1 + \epsilon_2$$

By using the double-angle identities of trigonometry, Eqs. (1.34) to (1.36) can be rewritten as

$$\epsilon_x' = \frac{\epsilon_x + \epsilon_y}{2} + \frac{\epsilon_x - \epsilon_y}{2} \cos 2\alpha + \frac{\gamma_{xy}}{2} \sin 2\alpha \qquad (1.34a)$$

$$\epsilon_y' = \frac{\epsilon_x + \epsilon_y}{2} - \frac{\epsilon_x - \epsilon_y}{2} \cos 2\alpha - \frac{\gamma_{xy}}{2} \sin 2\alpha \qquad (1.35a)$$

$$\frac{\gamma_{x'y'}}{2} = \frac{\epsilon_y - \epsilon_x}{2} \sin 2\alpha + \frac{\gamma_{xy}}{2} \cos 2\alpha \qquad (1.36a)$$

EXAMPLE 1.6

A strain gage is mounted on a specimen inclined at an angle θ with respect to the x axis, as shown in Fig. E1.6. When the specimen is subjected to a loading to produce $\epsilon_x = 800\mu\epsilon$, $\epsilon_y = 500\mu\epsilon$, and $\gamma_{xy} = -700\mu\epsilon$, the gage gives a reading of $411\mu\epsilon$. Determine the angle θ.

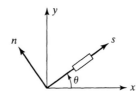

FIGURE E1.6

Solution

Since $\epsilon_n + \epsilon_s = \epsilon_x + \epsilon_y$,

$$\epsilon_n = \epsilon_x + \epsilon_y - \epsilon_s = 800 + 500 - 411 = 889\mu\epsilon$$

Subtracting Eq. (1.35a) from Eq. (1.34a), with $\alpha = \theta$, gives

$$\epsilon_s - \epsilon_n = (\epsilon_x - \epsilon_y) \cos 2\theta + \gamma_{xy} \sin 2\theta$$
$$411 - 889 = (800 - 500)[(1 - \sin^2 2\theta)^{1/2}] + (-700) \sin 2\theta$$

Rearranging the equation above gives

$$\sin^2 2\theta - 1.1538 \sin 2\theta + 0.2388 = 0$$
$$\sin 2\theta = \frac{1.1538 \pm [(-1.1538)^2 - 4(0.2388)]^{1/2}}{2}$$

When $\sin 2\theta = 0.8835$,

$$\theta = 31.0° \quad \text{or} \quad 59.0°$$

When $\sin 2\theta = 0.2703$,

$$\theta = 7.84° \quad \text{or} \quad 82.16°$$

Check: When $\theta = 31.0°$,

$$\epsilon_s = 800 \cos^2 31° + 500 \sin^2 31° - 700 \sin 31° \cos 31° = 411\mu\epsilon$$

When $\theta = 59.0°$,

$$\epsilon_s = 800 \cos^2 59° + 500 \sin^2 59° - 700 \sin 59° \cos 59° = 271\mu\epsilon$$

Comparing with the gage reading of $411\mu\epsilon$, $\theta = 31.0°$ is the correct solution, but $\theta = 59.0°$ is a plausible but incorrect value. Similarly, it can be shown that $\theta = 82.16°$ is also a possible answer, but $\theta = 7.84°$ is incorrect. Since $31°$ and $82.16°$ are quite different, the actual answer can be determined easily by inspection of the gage position.

A comparison of Eqs. (1.34a) to (1.36a) with Eqs. (1.7a) to (1.9a) shows that if we put

$$\sigma_x = \epsilon_x$$
$$\sigma_y = \epsilon_y$$
$$\tau_{xy} = \frac{\gamma_{xy}}{2}$$

the equations are of identical form. This transformation may be applied to all the previously derived stress equations to obtain the corresponding strain equations. Thus pertinent equations are written below without further derivations. The principal strains, the principal angle, and the maximum shear strain can be obtained by

$$\epsilon_1 = \epsilon_{max} = \frac{\epsilon_x + \epsilon_y}{2} + \left[\left(\frac{\epsilon_x - \epsilon_y}{2} \right)^2 + \left(\frac{\gamma_{xy}}{2} \right)^2 \right]^{1/2}$$

$$\epsilon_2 = \epsilon_{min} = \frac{\epsilon_x + \epsilon_y}{2} - \left[\left(\frac{\epsilon_x - \epsilon_y}{2} \right)^2 + \left(\frac{\gamma_{xy}}{2} \right)^2 \right]^{1/2} \tag{1.37}$$

$$\tan 2\alpha = \frac{\gamma_{xy}}{\epsilon_x - \epsilon_y} \tag{1.38a}$$

where $0 < \alpha < \pi/2$ if $\gamma_{xy} > 0$, and $-\pi/2 < \alpha < 0$ or $\pi/2 < \alpha < \pi$ if $\gamma_{xy} < 0$; or

$$\alpha_i = \tan^{-1} \frac{2(\epsilon_i - \epsilon_x)}{\gamma_{xy}} = \tan^{-1} \frac{\gamma_{xy}}{2(\epsilon_i - \epsilon_y)} \quad (i = 1, 2) \tag{1.38b}$$

$$\frac{\gamma_{max}}{2} = \left[\left(\frac{\epsilon_x - \epsilon_y}{2} \right)^2 + \left(\frac{\gamma_{xy}}{2} \right)^2 \right]^{1/2} = \frac{\epsilon_1 - \epsilon_2}{2} \tag{1.39}$$

EXAMPLE 1.7

Determine the principal strains and the principal angle for the given strain state at a point $\epsilon_x = 80\mu\epsilon$, $\epsilon_y = 200\mu\epsilon$, and $\gamma_{xy} = -160\mu\epsilon$. The x and y axes are horizontal and vertical, respectively.

Solution
Using Eq. (1.37), the values of principal strains can be obtained as

$$\epsilon_1 = \frac{(80 + 200) + [(80 - 200)^2 + (-160)^2]^{1/2}}{2}$$

$$= 140 + 100 = 240\mu\epsilon$$
$$\epsilon_2 = 140 - 100 = 40\mu\epsilon$$

By Eq. (1.38a), $-\pi/2 < \alpha < 0$, since $\gamma_{xy} = -160\mu\epsilon < 0$. Thus

$$\alpha_2 = 1/2 \tan^{-1}\frac{-160}{80 - 200} = 26.6°$$
$$\alpha_1 = \alpha_2 - 90° = -63.4° \quad (\text{CW})$$

Or by Eq. (1.38b),

$$\alpha_1 = \tan^{-1}\frac{-160/2}{240 - 200} = \tan^{-1}(-2) = -63.4° \quad (\text{CW})$$

The equation for Mohr's strain circle becomes

$$\left(\epsilon - \frac{\epsilon_x + \epsilon_y}{2}\right)^2 + \left(\frac{\gamma}{2}\right)^2 = \left(\frac{\epsilon_x - \epsilon_y}{2}\right)^2 + \left(\frac{\gamma_{xy}}{2}\right)^2 \tag{1.40}$$

The procedures for determining strain by using Mohr's circle are exactly the same as those described for stress. The only necessary modification is that the vertical axis is $\gamma/2$ in Mohr's strain circle, whereas it is τ in Mohr's stress circle.

EXAMPLE 1.8

To illustrate the point above, let's solve the problem presented in Example 1.7, this time by using Mohr's strain circle method.

Solution
Step 1. Set up the ϵ–$\gamma/2$ coordinate system, as shown in Fig. E1.8, where one division is chosen to represent $40\mu\epsilon$ for both the ϵ and $\gamma/2$ axes. Note that the scale must be the same for both axes. Also, the use of engineering paper is recommended.

Step 2. Locate the points that represent $(\epsilon_x, -\gamma_{xy}/2)$ and $(\epsilon_y, \gamma_{xy}/2)$. According to the sign conventions, point x is at $(80\mu\epsilon, 80\mu\epsilon)$ and point y is at $(200\mu\epsilon, -80\mu\epsilon)$, as shown in Fig. E1.8.

Step 3. Connect points x and y with a line, which intersects the ϵ axis at point o. With the center at o and with a radius of ox (or oy), draw a circle, as shown in Fig. E1.8. This is the Mohr's strain circle.

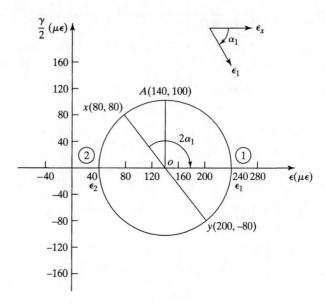

FIGURE E1.8   Mohr's strain circle.

Step 4. Analysis of Mohr's strain circle
(a) Points 1 and 2 are principal planes. The strain at point 1 is the maximum principal strain ϵ_1. Figure E1.8 indicates that ϵ_1 is approximately equal to 240$\mu\epsilon$ (6 divisions). Similarly, ϵ_2 is approximately equal to 40$\mu\epsilon$ (1 division). Note that only approximate values can be obtained by this graphical method.
(b) The angle between $o x$ and $o1$ is approximately 127°, which is twice that of the principal angle. The principal angle α equals –63.5° (clockwise direction).
(c) Point A represents the maximum shear strain (actually, $\gamma_{max}/2$). The magnitude of the maximum shear strain is equal to twice the radius, which is approximately 200$\mu\epsilon$. In summary, ϵ_1= 240$\mu\epsilon$, ϵ_2= 40$\mu\epsilon$, α = –63.5°, and γ_{max} = 200$\mu\epsilon$.

1.9   STRESS–STRAIN RELATIONS

For most engineering materials subjected to a load, it can be observed experimentally that the amount of deformation is somewhat proportional to the applied load within certain stress limits, namely,

$$\epsilon_x = \frac{\sigma_x}{E} \tag{1.41}$$

where E is called *Young's modulus* or the *modulus of elasticity,* which is the slope of the straight-line portion of a stress–strain curve in a uniaxial tension test. Equation (1.41) is well known as *Hooke's law,* one of the basic assumptions employed in the theory of linear elasticity.

Experimental evidence also shows that lateral strains ϵ_y and ϵ_z with opposite signs will be produced at the same time by the uniaxial stress σ_x, namely,

$$\epsilon_y = \epsilon_z = -\mu\, \frac{\sigma_x}{E} = -\mu\epsilon_x \qquad (1.42)$$

where the constant μ relating these strains is called *Poisson's ratio,* which has a value close to $\frac{1}{3}$ for most structural metals.

Since only linear elastic deformations are considered here, the well-known *principle of superposition* can be employed. This principle states that two or more strain (or stress) fields may be combined by direct superposition and the order of the application does not affect the final strain (or stress) field under consideration. By superposition of effects caused by σ_x, σ_y, and σ_z, Hooke's law can be generalized as follows:

$$\epsilon_x = \frac{\sigma_x}{E} - \mu\, \frac{\sigma_y}{E} - \mu\, \frac{\sigma_z}{E}$$
$$\epsilon_y = -\mu\, \frac{\sigma_x}{E} + \frac{\sigma_y}{E} - \mu\, \frac{\sigma_z}{E} \qquad (1.43)$$
$$\epsilon_z = -\mu\, \frac{\sigma_x}{E} - \mu\, \frac{\sigma_y}{E} + \frac{\sigma_z}{E}$$

The first, second, and third columns in Eq. (1.43) represent strains due to stresses σ_x, σ_y, and σ_z, respectively. It is found experimentally that the shear stresses and shear strains are also linearly related; that is,

$$\gamma_{xy} = \frac{\tau_{xy}}{G}$$
$$\gamma_{yz} = \frac{\tau_{yz}}{G} \qquad (1.44)$$
$$\gamma_{zx} = \frac{\tau_{zx}}{G}$$

where G is the modulus of rigidity or modulus of elasticity in shear.

Equations (1.43) and (1.44) represent the generalized Hooke's law for an isotropic solid subjected to three-dimensional loadings. It can be shown that Young's modulus E, the modulus of rigidity G, and Poisson's ratio μ are related by

$$G = \frac{E}{2(1 + \mu)} \qquad (1.45)$$

Most problems pertaining to experimental stress analysis aim to determine strains and stresses. They are performed either in a flat platelike structure whose thickness is small compared with its other dimensions when subjected to a load system in its own plane or on a free surface of a more complicated three-dimensional body. In both cases, no stresses act in the normal direction of the particular areas being investigated. These are the two-dimensional cases of what is called the *plane stress.* The general conditions required for plane stress are that $\sigma_z = \tau_{xz} = \tau_{yz} = 0$. Hooke's law in plane stress then becomes

$$\epsilon_x = \frac{1}{E}(\sigma_x - \mu\sigma_y)$$

$$\epsilon_y = \frac{1}{E}(\sigma_y - \mu\sigma_x) \tag{1.46}$$

$$\gamma_{xy} = \frac{\tau_{xy}}{G}$$

or solved for stresses,

$$\sigma_x = \frac{E}{1-\mu^2}(\epsilon_x + \mu\epsilon_y)$$

$$\sigma_y = \frac{E}{1-\mu^2}(\epsilon_y + \mu\epsilon_x) \tag{1.47}$$

$$\sigma_z = 0 \qquad \tau_{xy} = G\gamma_{xy}$$

The other particular case in two dimensions is called *plane strain*. Whereas $\sigma_z = 0$ in plane stress, $\epsilon_z = 0$ in plane strain, so that $\sigma_z = \mu(\sigma_x + \sigma_y)$. Plane strain is less important in the experimental stress analysis and is not considered here.

Previously, the compatibility equation in two dimensions, Eq. (1.32), was expressed in terms of strains. Sometimes it is more useful to express the compatibility equation in terms of stresses. Substituting Hooke's law in Eq. (1.46) into Eq. (1.32) and using Eq. (1.45) yields

$$2(1+\mu)\frac{\partial^2\tau_{xy}}{\partial x\,\partial y} = \frac{\partial^2(\sigma_x - \mu\sigma_y)}{\partial y^2} + \frac{\partial^2(\sigma_y - \mu\sigma_x)}{\partial x^2} \tag{1.48}$$

If the body forces are neglected, the equilibrium equation in two dimensions becomes

$$\frac{\partial\sigma_x}{\partial x} + \frac{\partial\tau_{xy}}{\partial y} = 0$$

$$\frac{\partial\sigma_y}{\partial y} + \frac{\partial\tau_{xy}}{\partial x} = 0 \tag{1.21}$$

Differentiating the first equation of Eq. (1.21) with respect to x and the second equation of (1.21) with respect to y and then adding them together gives

$$\frac{\partial^2\sigma_x}{\partial x^2} + \frac{\partial^2\sigma_y}{\partial y^2} + 2\frac{\partial^2\tau_{xy}}{\partial x\,\partial y} = 0 \tag{1.49}$$

Solving Eqs. (1.48) and (1.49) to eliminate the derivative of τ_{xy} and then rearranging the terms yields

$$\left(\frac{\partial^2}{\partial x^2} + \frac{\partial^2}{\partial y^2}\right)(\sigma_x + \sigma_y) = \left(\frac{\partial^2}{\partial x^2} + \frac{\partial^2}{\partial y^2}\right)(\sigma_1 + \sigma_2) = 0 \tag{1.50}$$

This is the compatibility equation for plane stress and is known as the *Laplace equation in plane stress*.

If the body forces are not negligible, the compatibility equation becomes

$$\left(\frac{\partial^2}{\partial x^2} + \frac{\partial^2}{\partial y^2}\right)(\sigma_x + \sigma_y) = -(1 + \mu)\left(\frac{\partial X}{\partial x} + \frac{\partial Y}{\partial y}\right) \tag{1.51}$$

All required equations presented thus far have been derived to solve a problem in two dimensions. There are eight independent unknowns in two dimensions (i.e., three stresses: σ_x, σ_y, and τ_{xy}; three strains: ϵ_x, ϵ_y, and γ_{xy}, and two displacements: u and v). Now eight independent equations, two differential equations of equilibrium [Eqs. (1.19) and (1.20)], three stress–strain relations or Hooke's law [Eq. (1.47)], and three strain–displacement relations [Eq. (1.31)] have been derived so that any given problem is theoretically determinate.

However, it is sometimes desirable to work entirely in terms of strain and not to work in terms of both strains and displacements. The compatibility equation [Eq. (1.32) or Eq. (1.50)] must be used to ensure that the strains are compatible with single-valued displacements. In such cases, there are six independent unknowns in two dimensions: σ_x, σ_y, τ_{xy}, ϵ_x, ϵ_y, and γ_{xy}. The six independent equations to be used are two differential equations of equilibrium [Eqs. (1.19) and (1.20)], three stress–strain relations or Hooke's law [Eq. (1.47)], and one compatibility equation [Eq. (1.32) or Eq. (1.50)].

In this chapter we have reviewed basic equations in two dimensions. For more detailed derivations or more advanced formulations, the reader is referred to references [1] to [4].

PROBLEMS

1.1. The stresses at a point are shown in Fig. P1.1a. Determine stresses (σ, τ shown in Fig. P1.1b) on a plane that is inclined 30° CCW to the horizontal plane.

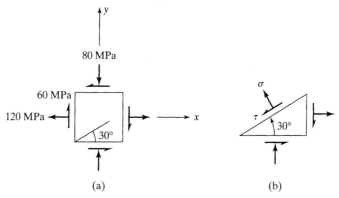

(a)   (b)

FIGURE P1.1

1.2. The state of plane stress at a point in a loaded body with respect to the xy coordinate axes is shown in Fig. P1.2a. Determine the stress state at the same point with respect to a new set of coordinate axes obtained by rotating the xy axes 60° CW, as shown in Fig. P1.2b. Check the invariance of the quantity I_1 (sum of the normal stresses).

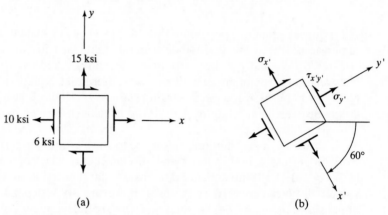

(a)   (b)

FIGURE P1.2

1.3. The stresses at a point with respect to the xy coordinate system are shown in Fig. P1.3. If one of the stress components at the same point with respect to the $x'y'$ coordinate system is known (i.e., $\sigma_{x'} = 18$ ksi), determine the other stress components $\sigma_{y'}$ and $\tau_{x'y'}$ and the angle α with respect to the x axis.

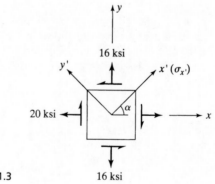

FIGURE P1.3   16 ksi

1.4. Prove that $\alpha = -64.74°$ and $\alpha = 38.18°$ are not correct answers in Example 1.2.

1.5. Verify Eq. (1.13).

1.6. The stress state at a point with respect to the xy-coordinate system is given as follows: $\sigma_x = 120$ MPa, $\sigma_y = 60$ MPa, $\tau_{xy} = -40$ MPa. Determine: **(a)** the principal stresses and principal angle, and **(b)** the maximum shear stress. Sketch the given state of stress and the stress states in parts (a) and (b) on suitably oriented stress elements.

1.7. Verify Eq. (1.15).

1.8. The stress state at a point is given by $\sigma_x = -14$ ksi, $\sigma_y = 16$ ksi, $\tau_{xy} = -12$ ksi. Using Mohr's stress circle method, determine **(a)** the principal stresses and principal angle, **(b)** the maximum shear stress, and **(c)** the stresses on a plane whose normal is at 45° CW with the x axis. Sketch the stress state.

1.9. An arch (a quarter circle) in a plate is shown in Fig. P1.9. The plate is subjected to a uniform strain loading (ϵ_x, ϵ_y, and γ_{xy} are constant everywhere). Determine the length change Δl of the arch.

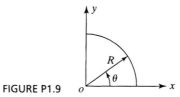

FIGURE P1.9

1.10. The state of plane strain at a point in a loaded member is given as $\epsilon_x = -550\mu\epsilon$, $\epsilon_y = 1100\mu\epsilon$, $\gamma_{xy} = 1800\mu\epsilon$. Determine **(a)** the maximum shear strain, and **(b)** the principal strains and principal angle.

1.11. To determine the strain state in a structural member assumed to be in a uniform strain field, three single strain gages are mounted on the structural member, as shown in Fig. P1.11. At the operating load, the gage readings are $\epsilon_a = 1500\mu\epsilon$, $\epsilon_b = 1000\mu\epsilon$ and $\epsilon_c = -500\mu\epsilon$. Determine the principal strains and principal angle.

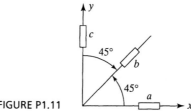

FIGURE P1.11

1.12. The strain state at a point is given as $\epsilon_x = 900\mu\epsilon$, $\epsilon_y = 100\mu\epsilon$, $\gamma_{xy} = -600\mu\epsilon$. Using Mohr's strain circle method, determine **(a)** the maximum shear strain, **(b)** the principal strains and principal angle, and **(c)** the strain components with respect to the $x'y'$ coordinate system, achieved by rotating the x axis 45° CCW.

REFERENCES

[1] A. S. Khan and M. Malik, *Elementary Solid Mechanics,* in review.

[2] A. J. Durelli, E. Phillips, and C. Tsao. *Introduction to the Theoretical and Experimental Analysis of Stress and Strain,* McGraw-Hill, New York, 1958.

[3] S. P. Timoshenko and J. N. Goodier, *Theory of Elasticity,* 2nd ed., McGraw-Hill, New York, 1951.

[4] R. J. Roark and W. C. Young, *Formulas for Stress and Strain,* 5th ed., McGraw-Hill, New York, 1975.

[5] R. E. Peterson, *Stress Concentration Factors; Charts and Relations Useful in Making Strength Calculations for Machine Parts and Structural Elements,* Wiley, New York, 1974.

CHAPTER 2

Metal-Foil Resistance Strain Gages

2.1 INTRODUCTION

Experimental stress analysis is a misnomer because strain is normally measured and stress is then calculated by using stress–strain relationships. Perhaps the most common measurement is the surface deformation between two points or along a length l_0, called the *gage length*. Any device that is used to measure surface deformation can be classified as a strain gage. Normally, gages are categorized by their construction into four groups: mechanical, optical, electrical, and acoustical. Among them, the most important and widely used gage is the electrical-resistance type, where small changes in dimension result in equivalent changes in resistance. This type of strain gage is used for 80% or more of the industrial experimental stress analyses performed in the United States today [1]. Therefore, the electrical-resistance strain gage is covered in detail in this and following chapters.

2.2 PRINCIPLE OF OPERATION

The operative principle of the electrical-resistance strain gage has been known for more than a century. In 1856, Lord Kelvin [2] reported that certain metal wires exhibited a "change of electrical resistance with change in strain." The total electrical resistance of a rectangular uniform-cross-section conductor is given by the equation

$$R = \frac{rL}{ab} \tag{2.1}$$

where $R, r, L, a,$ and b are the resistance, specific resistance, length, and lateral dimensions of the rectangular cross section of the conductor, respectively. Taking logarithms and differentiating Eq. 2.1 leads to

$$\frac{dR}{R} = \frac{dr}{r} + \frac{dL}{L} - \frac{da}{a} - \frac{db}{b} \tag{2.2}$$

For an axial strain, $\epsilon_a = dL/L$, there is a transverse strain ϵ_t, which is given by

$$\epsilon_t = -\mu\epsilon_a = -\mu\frac{dL}{L} = \frac{-da}{a} = \frac{-db}{b}$$

where μ is Poisson's ratio. Substituting the equation above into Eq. (2.2) gives

$$\frac{dR}{R} = \frac{dr}{r} + (1 + 2\mu)\frac{dL}{L} \tag{2.3}$$

Assuming that the change in length and change in specific resistance are small, the higher-order terms can be neglected. Thus

$$\frac{\Delta R}{R} \approx \frac{\Delta r}{r} + (1 + 2\mu)\frac{\Delta L}{L} \tag{2.4}$$

To describe the electrical resistance change of a conductor caused by the change of its length, the term *strain sensitivity* is introduced. It is defined as the resistance change (ΔR) per unit of initial resistance (R) per unit of applied axial strain. Strain sensitivity is denoted by S_a. By definition,

$$S_a = \frac{\Delta R/R}{\Delta L/L} = \frac{\Delta R/R}{\epsilon_a} \tag{2.5}$$

where ϵ_a is the nominal strain in the axial direction.

By using Eq. (2.4), Eq. (2.5) can be rewritten as

$$S_a = \frac{\Delta r/r}{\epsilon_a} + (1 + 2\mu) \tag{2.6}$$

From Eq. (2.6) we find that the strain sensitivity of a resistance element is produced by two factors: the term $(\Delta r/r)/\epsilon_a$ denoting the change in specific resistance of the conductor material, and the term $(1 + 2\mu)$ representing the change in the dimensions of the conductor. Since Poisson's ratio is approximately 0.3 for most metal alloys used as the resistance element, the strain sensitivity will be about 1.6 if only the dimensional changes are considered.

In some cases (e.g., semiconductor gages) the specific resistance term dr/r is much more dominant than other terms, and the contribution from the term $(1 + 2\mu)(dL/L)$ in Eq. (2.3) is usually small and negligible. For large strain, the resistance element undergoes plastic deformation. Under the usual assumptions that there is no plastic volume change and that dr/r is approximately equal to dv/v for foil of copper and nickel, the commonly used gage materials, Eq. (2.3) reduces to

$$\frac{dR}{R} = 2\frac{dL}{L} \tag{2.7}$$

since $dv/v = (1 - 2\mu)\,dL/L$. Integration of Eq. (2.7) results in

$$\ln \frac{R}{R_0} = 2 \ln \frac{L}{L_0}$$

or

$$\ln \left(1 + \frac{\Delta R}{R_0} \right) = 2 \ln \left(1 + \frac{\Delta L}{L_0} \right)$$

Therefore,

$$\frac{\Delta R}{R_0} = 2 \frac{\Delta L}{L_0} + \left(\frac{\Delta L}{L_0} \right)^2$$

Substituting the equation above into Eq. (2.5), we obtain

$$S_a = 2 + \epsilon_a \tag{2.8}$$

where $\epsilon_a = \Delta L/L_0$ is the nominal strain. Although the assumptions used in the derivations of Eq. (2.8) have yet to be verified for metal-foil gages, it is well known and widely used in practice for large strain measurements [16].

Table 2.1 shows some typical values of strain sensitivity during elastic deformation for those metallic alloys commonly used in the manufacture of commercially available strain gages [3,11–13]. Note that S_a varies from 2.0 to 3.6 for these common alloys. For isoelastic, the specific resistance term $(\Delta r/r_0)/\epsilon_a$ is a significant contributor to strain sensitivity, since the term $(1 + 2\mu)$ is approximately equal to 1.6.

Constantan or annealed Constantan [(45% Ni, 55% Cu) or (40% Ni, 60% Cu)] is the alloy commonly used for general-purpose strain gages. It has several advantages. First, its strain sensitivity is high and relatively insensitive to strain level over a very wide range of strain (up to 8%), useful to measure both elastic as well as plastic strain in many structural materials. Second, its resistivity is high, so it is possible to construct a small gage with a relatively high resistance. Third, temperature changes do not have a significant effect when used on common structural materials due to its excellent thermal stability. Finally, the ability to control the small temperature-induced changes in

TABLE 2.1   Strain Sensitivity S_a for Common Strain Gage Alloys

| Material | S_a | Manufacturer Designation | | r $\mu\Omega$ – ft $(\mu\Omega \cdot \text{m})$ |
		Micro-Measurements	Kyowa	
Constantan	2.1	A	FD	0.15 (0.49)
Annealed constantan	2.1	P	FE	0.15 (0.49)
Karma	2.0	K		0.41 (1.35)
Nichrome V	2.1	K	H	0.30 (0.98)
Isoelastic	3.6	D		0.34 (1.12)

resistance with trace impurities or by heat treatment makes this alloy very versatile in fabricating temperature-compensated strain gages. The thermal expansion coefficient of the gage alloy can be matched with that of many different engineering materials. The isoelastic (36% Ni, 55.5% Fe, 8% Cr, 0.5% Mo) and the karma (74% Ni, 20% Cr, 3% Al, 3% Fe) alloys are widely used for special purpose gages. For example, the isoelastic alloy is used in dynamic and fatigue applications, due to its high sensitivity and high fatigue strength, and is also used in special-purpose transducers to meet the high output requirement. The karma alloy is used in temperature-compensated gages to achieve the compensation over a larger range of temperature and is preferred for accurate static measurement over long periods, due to its excellent stability. Nichrome (80% Ni, 20% Cr) alloy is used in very special applications where high temperatures (up to +750°F or +400°C) are involved [13].

2.3   GAGE CONFIGURATIONS AND FABRICATION

The common electrical-resistance strain gages used universally today are bonded gages, which consist of a strain sensing element, a thin film backing that serves both as an insulator and as a carrier for the strain sensing element, and terminals for lead wire connections. A strain sensing element may consist of either a length of very fine wire, 0.025 mm (0.001 in.) or less in diameter [17], looped into a grid pattern to produce the necessary length for a specific resistance value, or a grid that is photoetched from a very thin sheet of metallic foil. The former is called the *bonded-wire strain gage,* named SR-4 to honor the people deemed most responsible for its development [6]. The latter is called the *bonded-metal-foil strain gage.* Typical SR-4 gages are shown in Fig. 2.1.

For gages about 25 mm (1 in.) or less long, a flat grid is usually adopted. For very short gage lengths, the gages are usually wound in a porous bobbin card [17]. Bonded-wire strain gages are used in only a few applications today, having largely been replaced by bonded-metal-foil strain gages. The earliest metal-foil strain gages were manufactured by Saunders and Roe in England in 1952 [3]. Since then, a variety of product lines with rigorously controlled specifications have been established by many strain gage manufacturers to produce various high-quality precision strain gages.

Figure 2.2 displays some of the metal-foil strain gages currently available commercially. Standard gage resistances are 120 Ω and 350 Ω. Commercially available gage

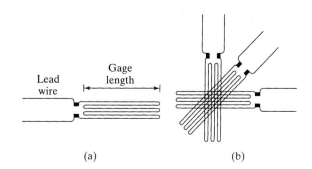

FIGURE 2.1   Typical SR-4 wire gages: (a) flat grid; (b) rosette [4].

(a)   (b)

FIGURE 2.2 Configurations of selected metal-foil gages (Micro-Measurements Division of Measurements Group, Hartrun Corp., Kyowa, and Brüel & Kjaer).

lengths range from 0.20 mm (0.008 in.) to 100 mm (4 in.) [17] to meet a variety of strain gage applications. The gages shown in Fig. 2.2a to c, called *single-element gages,* are used to measure strain in one direction only. However, when coupled with different grid configurations or multigrid arrangements, as shown in Fig. 2.2d to i, improved measurements or additional information can be obtained with little additional effort.

Figures 2.2d and Fig. 2.2e illustrate *two-element rectangular rosettes* in planar and stacked configurations, respectively, for applications where the principal directions are known. By using these rosettes, principal strains can be determined by aligning each element of the rosette along known principal directions. *Three-element rosettes* in either a *rectangular (45°) orientation,* shown in Fig. 2.2f and g, or a *delta (60°) orientation,* shown in Fig. 2.2h and i, are available for general two-dimensional applications where the principal directions as well as the magnitudes of the principal strains are unknown. The stacked arrangement is adopted to save space and to give a closer approximation to actual strain measurement at a point. It should be noted, however, that heat dissipation can be a problem when this type of rosette is used with higher excitation voltages.

Several special-purpose gages are shown in Fig. 2.2j to r. Those illustrated in Fig. 2.2j to l are used to measure tangential, radial, or combined strains on thin membranes and diaphragms. The gages shown in Fig. 2.2m and n are frequently employed to measure torsional strains on axles and shafts. With the symmetric axis of the gage mounted along the axis of the shafts, this rosette provides readings for shear strain, γ_{xy}, under application of a twisting moment. It can also be calibrated to give readings in torque and is even sometimes called the *torque gage.*

The gages shown in Fig. 2.2o and p are called *stress gages* because they can be calibrated to provide readings directly for stress. Recall that in a plane stress state ($\sigma_z = 0$), σ_x and σ_y are given by Eq. (1.47) as follows:

$$\sigma_x = \frac{E}{1 - \mu^2} (\epsilon_x + \mu\epsilon_y)$$

$$\sigma_y = \frac{E}{1 - \mu^2} (\epsilon_y + \mu\epsilon_x)$$

where E is Young's modulus and μ is Poisson's ratio. The rosette illustrated in Fig. 2.2o has two grids with lengths in the ratio $1{:}\mu$. In a particular application, the longer grid is aligned in the same direction as that of the required stress measurement (say, in the σ_x direction). Under the application of load, the two grids connected in series give the ($\epsilon_x + \mu\epsilon_y$) term in the equation above, so that the output of the rosette is directly proportional to stress σ_x. The gage shown in Fig. 2.2p is used in a slightly different manner. The ratio of the components of the gage conductors in the direction of the measurement and at its right angle is $1{:}\mu$. The two elements connected in series also give the ($\epsilon_x + \mu\epsilon_y$) term when the central axis of the gage is oriented in the direction of the desired measurement (σ_x). Finally, the gages shown in Fig. 2.2q and r are called *crack propagation gages.* The gages change in resistance either discontinuously (Fig. 2.2q) or "continuously" (Fig. 2.2r), as the growing crack will progressively break the gage conductors, as in the former case, or reduce the conducting area of cross section, as in the latter case. This is an example of nonstrain measurement application.

EXAMPLE 2.1

A stress gage (Fig. 2.2p) is bonded to a specimen, shown in Fig. E2.1. Determine the angle ϕ in terms of Poisson's ratio μ. Also derive the expression $\Delta R/R$ in terms of S_a, E, and σ_x, where S_a is the strain sensitivity, and E is Young's modulus of the test material, respectively. Assume that the $\Delta R/R$ of the gage is the same as that of a single conductor.

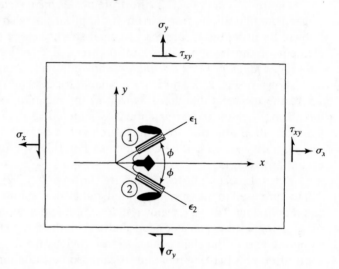

FIGURE E2.1

Solution   The stress gage is considered as two gages connected in series. Using Eq. (1.34a), we have

$$\epsilon_1 = \frac{\epsilon_x + \epsilon_y}{2} + \frac{\epsilon_x - \epsilon_y}{2}\cos 2\phi + \frac{\gamma_{xy}}{2}\sin 2\phi$$

$$\epsilon_2 = \frac{\epsilon_x + \epsilon_y}{2} + \frac{\epsilon_x - \epsilon_y}{2}\cos 2(-\phi) + \frac{\gamma_{xy}}{2}\sin 2(-\phi)$$

$$= \frac{\epsilon_x + \epsilon_y}{2} + \frac{\epsilon_x - \epsilon_y}{2}\cos 2\phi - \frac{\gamma_{xy}}{2}\sin 2\phi$$

Thus the gage reads the average strain, namely,

$$\epsilon = 0.5(\epsilon_1 + \epsilon_2) = 0.5[(\epsilon_x + \epsilon_y) + (\epsilon_x - \epsilon_y)\cos 2\phi]$$

$$= 0.5[\epsilon_x(1 + \cos 2\phi) + \epsilon_y(1 - \cos 2\phi)]$$

$$\epsilon = \frac{1 + \cos 2\phi}{2}\left(\epsilon_x + \frac{1 - \cos 2\phi}{1 + \cos 2\phi}\epsilon_y\right)$$

For plane stress, Eq. (1.47) gives

$$\sigma_x = \frac{E}{1 - \mu^2}(\epsilon_x + \mu\epsilon_y) \quad \text{or} \quad \epsilon_x + \mu\epsilon_y = \frac{1 - \mu^2}{E}\sigma_x$$

Since $\Delta R/R$ of a gage is assumed the same as that of a single conductor, Eq. (2.5) can be employed. Thus

$$\frac{\Delta R}{R} = S_a\epsilon = S_a\frac{1 + \cos 2\phi}{2}\left(\epsilon_x + \frac{1 - \cos 2\phi}{1 + \cos 2\phi}\epsilon_y\right)$$

The angle ϕ is so chosen that

$$\frac{1 - \cos 2\phi}{1 + \cos 2\phi} = \mu \quad \text{or} \quad \tan^2\phi = \mu$$

Thus, $\phi = \tan^{-1}\sqrt{\mu}$. It can be shown that $1 + \cos 2\phi = 2/(1 + \mu)$. Therefore

$$\frac{\Delta R}{R} = \frac{S_a}{1 + \mu}(\epsilon_x + \mu\epsilon_y) = \frac{(1 - \mu)\,S_a}{E}\,\sigma_x$$

Note that were the gage to be mounted along the axis of the shaft and the two small gages connected in a half bridge (see Chapter 3), gage readings would give shear strain (or torque).

2.4 STRAIN GAGE BONDING AGENTS AND PROCEDURES

It is very important to employ the proper adhesive and bonding procedures to achieve precise strain measurements by using bonded resistance strain gages. Strain gage manufacturers provide several different adhesives specially formulated for a wide variety of strain applications. Selection of a proper adhesive is strongly dependent on the carrier material, operating and curing temperatures, and the maximum strain to be measured. The most commonly used bonding agents are methyl-2-cyanoacrylate, epoxy, and ceramic-based adhesives.

Micro-Measurements M-Bond 200 adhesive [18], for example, is a modified methyl-2-cyanoacrylate compound that is an excellent general-purpose laboratory adhesive. This adhesive has the advantages of fast room-temperature cure and ease of application. M-Bond 200 adhesive is normally used where measured strains are less than 3% or $30,000\mu\epsilon$, where temperatures are in the range $+200°F$ $(+95°C)$ to $-300°F$ $(-185°C)$, and in quasistatic as well as cycle loading applications.

For common strain gage applications, a thin layer of the M-Bond 200 adhesive is placed between the gage and the surface of the test sample or structure, and a gentle pressure is applied for at least 1 minute to induce polymerization. Although this adhesive requires neither heat nor a hardening agent to induce polymerization, an M-Bond catalyst, specially formulated to control the reactivity rate of this adhesive, is usually applied to decrease the reaction time. Adequate protective coatings are necessary when the measurement is undertaken in a high-humidity environment, because the adhesiveness of the M-Bond 200 becomes ineffective due to absorption of moisture. This bonding agent will generally become harder and more brittle with time or exposure to elevated temperatures. It is not recommended in applications for long-term use (e.g., more than one year).

For high-elongation strain measurements in excess of 3% $(30,000\mu\epsilon)$ but not exceeding 15% $(150,000\mu\epsilon)$, epoxy adhesives are employed. An epoxy bonding agent usually consists of two constituents, a resin and a curing agent. It should be noted that the amount of curing agent added to the resin is extremely important. The adhesive curing temperatures and the residual stresses produced during polymerization will be greatly influenced by as little as 1 or 2% variation of the amount from the specified values listed by the manufacturers. Therefore, the amounts of both the resin and the curing agent should be weighed carefully before they are mixed together. A pressure of 5 to 20 psi (or 34.5 to 138 kPa) is recommended for the epoxies during the cure period to ensure as thin a layer of adhesive as possible.

Many different epoxy systems in kit form are commercially available today. A wide variation in properties can be obtained with different resin and curing agent combinations. For example, the Micro-Measurements M-Bond AE-10 system [19] is composed of resin AE and the Curing Agent 10. If the adhesive is cured at 70°F (21°C) for 6 hours, it is capable of 6% elongation. When the curing time is extended to 24 to 48 hours at 75°F (24°C), elongation capabilities will be increased to 10% or even more (to 20% for an uniaxial compression test).

Ceramic-based bonding agents are used for high-temperature applications. Strain measurements at temperatures greater than 260°C (or 500°F) are very challenging. Interested readers are directed to Chapter 13 of *Handbook On Experimental Mechanics* [6] for further reading on this subject.

Proper bonding of a strain gage to a specimen is perhaps one of the most critical steps in the entire course of measuring strain with a bonded resistance strain gage. Usually, instructions are provided with the adhesive kit by the strain gage manufacturers to show the important steps to be followed to bond a strain gage. When mounting a gage on a specimen, it is very important to prepare carefully the surface of the specimen where the gage is to be bonded. The surface should be free of rust, scale, paint, and so on, and should be smooth but not highly polished. Also, the surface must be cleaned thoroughly by using solvents to remove all traces of oil or grease. It is a good policy to keep cleaning with cotton swabs or with a gauze sponge until it no longer picks up dirt. The gage location is then marked on the specimen by a 4H pencil or a ballpoint pen, depending on the hardness of the specimen material. Finally, the surface may be treated with a basic solution to produce proper chemical affinity for the bonding agent.

The grit of the sandpaper or silicon carbide paper used for surface cleaning usually depends on the hardness of the specimen material and the strain to be measured. Experience shows that the softer the specimen material, the higher the grit number, and the larger the strain to be measured, the lower the grit number. For example, 320-grit silicon carbide paper is normally used for aluminum specimens. Strokes used in cleaning with abrasive paper are oriented at ±45° to the intended axis of strain measurement. The gage is then positioned using cellophane tape, which will keep the gage in the correct position during the application of the bonding agent and subsequent squeezing to get rid of any excess adhesive. The application procedures for adhesives vary with the type of bonding agent and are detailed by the manufacturer.

After a strain gage has been bonded to the surface of a specimen and the adhesive has cured, lead wires are attached via intermediate anchor terminals, as shown in Fig. 2.3. For gages without leads (Fig. 2.3a), care should be taken in soldering leads to the soldering tabs of the gages and anchor terminals. A small-diameter wire approximately 1 in. long is preferred. Metal-foil strain gages are relatively fragile, and overheating can break down the adhesive or backing of the terminals. Whenever possible, gage resistance should be checked and recorded. If the resistance is quite different

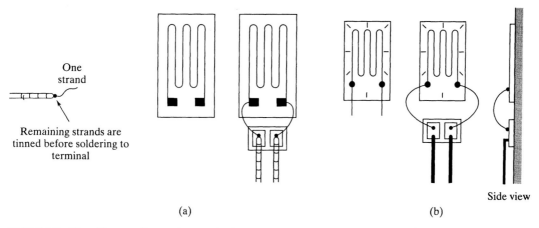

One
strand

Remaining strands are
tinned before soldering to
terminal

(a)

(b)

Side view

FIGURE 2.3   Use of intermediate anchor terminals to attach the lead wires to strain gages [9].

from the rated value, the gage should be replaced, as it would be very difficult to balance it in a strain indicator.

The gage insulation from the specimen should be checked using a megohm meter with low excitation voltage, since the use of high voltage may damage the gage or the adhesive bond. This resistance should register a minimum of 1000 MΩ [17]. This can also be done using a strain indicator. If the reading is unstable, the gage and/or terminal insulations are not good enough. To check whether a gage is firmly bonded, run a fingertip over the mounted gage. Any movement of the gage caused by improper bonding can cause a drastic change in the values indicated.

2.5   GAGE FACTOR AND TRANSVERSE SENSITIVITY CORRECTION

The term *gage factor*, denoted by G_f, is used to describe a gage's sensitivity to strain. It is defined mathematically as follows:

$$G_f = \frac{\Delta R/R}{\Delta L/L} = \frac{\Delta R/R}{\epsilon_a} \qquad \epsilon_a = \frac{\Delta R/R}{G_f} \qquad (2.9)$$

where ϵ_a is the nominal strain along the axial direction of the gage. The equation for gage factor corresponds to that of the strain sensitivity of a single conductor [Eqs. (2.9) and (2.5)]. However, the conductor in a gage is formed in a grid pattern to produce a short gage length and the required gage resistance for strain measurements. Due to this, the gage factor will be slightly different from the strain sensitivity value of the corresponding conductor because the end loops in the flat-grid SR-4 wire gages or the large width/thickness ratio for each gridline in metal-foil gages affect overall sensitivity.

Let a strain gage, bonded on a specimen surface, be subjected to an axial strain ϵ_a, a transverse strain ϵ_t, and a shear strain. The relative resistance change of the gage $(\Delta R/R)$ does not appreciably depend on the shear strain and thus can be expressed as

$$\frac{\Delta R}{R} = S_{ga}\epsilon_a + S_{gt}\epsilon_t \tag{2.10}$$

where S_{ga} is the gage sensitivity to the axial strain and S_{gt} is the gage sensitivity to the transverse strain. Equation (2.10) can be rewritten as

$$\frac{\Delta R}{R} = S_{ga}(\epsilon_a + K_t\epsilon_t) \tag{2.11}$$

or

$$\frac{\Delta R}{R} = S_{ga}\left(1 + \frac{K_t\epsilon_t}{\epsilon_a}\right)\epsilon_a \tag{2.12}$$

where $K_t = S_{gt}/S_{ga}$ is the transverse sensitivity factor of the gage. Comparing Eq. (2.12) with Eq. (2.9), we find that

$$G_f = S_{ga}\left(1 + \frac{K_t\epsilon_t}{\epsilon_a}\right) \tag{2.13}$$

Thus the gage factor depends on the measured strain field. The constant value for the gage factor, which is supplied by the strain gage manufacturer, is obtained by subjecting the gage to an uniaxial stress field. According to the American Society for Testing and Materials Standard E251, there are three ways to determine the gage factor of electrical-resistance strain gages at a reference temperature. They include the constant bending moment beam method, cantilever beam method, and direct tension and compression method. In all three cases, the gage is subjected to an uniaxial stress field, and the transverse and axial strains are related through Poisson's ratio of the material (μ_c). Thus

$$G_f = S_{ga}(1 - \mu_c K_t) \tag{2.14}$$

where μ_c is Poisson's ratio of the calibration beam or bar. In practice, the strain ϵ_a is calibrated at a given load by an extensometer. The sample gage is mounted on the calibrated beam. After the load is applied, the gage resistance change ΔR is measured and the gage factor G_f is determined by using Eq. (2.9). Usually, at least five measurements are obtained by mounting five identical strain gages on the calibrated beam and taking the average value as the gage factor for this type of gage.

EXAMPLE 2.2

To determine the gage factor, a strain gage is mounted on the lower surface of a beam as shown in Fig. E2.2. The initial resistance of the gage is 120.1 Ω. At a load P of 6745 lb, the resistance of the gage is increased to 120.8 Ω. What is the gage factor? Assume that $E = 30,000$ ksi, $L = 4.0$ in., $L_1 = 10.0$ in., $b = 0.5$ in., and $h = 2$ in.

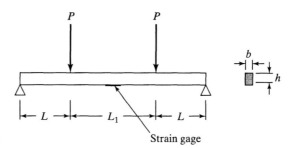

FIGURE E2.2   Strain gage

Solution   In the middle portion of the beam, the bending moment is constant,

$$M = PL = (6745)(4.0) = 26{,}980 \text{ in.-lb}$$

At the lower surface of the beam, the strain ϵ is given by

$$\epsilon = \frac{6M}{Ebh^2} = \frac{(6)(26{,}980)}{(30 \times 10^6)(0.5)(2^2)} = 0.002698$$

The resistance change of the gage, ΔR, is

$$\Delta R = 120.8 - 120.1 = 0.7 \ \Omega$$

The gage factor is obtained using Eq. (2.9):

$$G_f = \frac{0.7/120.1}{0.002698} = 2.16$$

EXAMPLE 2.3

A strain gage of 120 Ω with a gage factor of 2.04 is mounted on a tensile specimen. At a certain load, the gage resistance changes by 0.087 Ω. What is the strain in the direction of the gage axis?

Solution   By using Eq. (2.9), the strain ϵ_a in the direction of the gage axis is given by

$$\epsilon_a = \frac{\Delta R/R}{G_f} = \frac{0.087/120}{2.04} = 0.0003553 = 355.3\mu\epsilon$$

Table 2.2 shows some typical nominal values of G_f and K_t for several different kinds of gages manufactured by Measurements Group Inc. and Kyowa Electronic Instruments Company (Japan). Note that the values given in the table may vary ($\pm5\%$) from lot to lot of foil strain gage. The actual values are supplied with the strain gages by the manufacturers.

It should be noted that errors will be introduced in a strain gage measurement when Eq. (2.9) is used, except for two special cases: (1) when the transverse sensitivity K_t for the gage is zero, or (2) when the ratio of $-\epsilon_t/\epsilon_a$ is μ_c. In some instances the error may be negligible. In other cases, however, it is significant and corrections are neces-

TABLE 2.2 Gage Factor G_f and Transverse Sensitivity K_t

Gage Type	G_f	$K_t (\%)$
EA-06-250BG-120	2.05	0.3
EA-06-500AF-120	2.05	0.2
WA-06-250BG-120	2.05	−1.0
WK-06-250BG-350	2.00	−4.0
WK-06-500AF-350	2.00	−8.0
KFE-2-C1	2.12	0.4
KFC-1-C1	2.11	1.2
KFC-30-C1	2.11	3.0

Source: Data from Measurements Group and Kyowa.

sary during data analysis. The magnitude of the error can be estimated using the following equation:

$$\text{error} = \frac{\epsilon_a' - \epsilon_a}{\epsilon_a}(100) = \frac{K_t(\epsilon_t/\epsilon_a + \mu_c)}{1 - \mu_c K_t}(100) \tag{2.15}$$

where ϵ_a' is the indicated strain, obtainable if the gage factor given by the manufacturer is used, that is,

$$\epsilon_a' = \frac{\Delta R/R}{G_f} \tag{a}$$

From Eq. (2.15) it is clear that the error will be zero if either K_t is zero or $\epsilon_t/\epsilon_a = -\mu_c$, as mentioned earlier.

To derive Eq. (2.15), substitute Eq. (2.14) into Eq. (2.12):

$$\frac{\Delta R}{R} = \frac{G_f \epsilon_a}{1 - \mu_c K_t}\left(1 + K_t\frac{\epsilon_t}{\epsilon_a}\right) \tag{b}$$

or

$$\epsilon_a = \frac{\Delta R/R}{G_f}\frac{1 - \mu_c K_t}{1 + K_t\epsilon_t/\epsilon_a} \tag{c}$$

By using Eq. (a), Eq. (c) can be expressed as

$$\epsilon_a = \epsilon_a'\frac{1 - \mu_c K_t}{1 + K_t\epsilon_t/\epsilon_a} \tag{2.16}$$

Therefore,

$$\text{error} = \frac{\epsilon_a' - \epsilon_a}{\epsilon_a}(100) = \frac{\epsilon_a' - \epsilon_a'[(1 - \mu_c K_t)/(1 + K_t\epsilon_t/\epsilon_a)]}{\epsilon_a'[(1 - \mu_c K_t)/(1 + K_t\epsilon_t/\epsilon_a)]}(100)$$

$$= \frac{1 + K_t(\epsilon_t/\epsilon_a) - 1 + \mu_c K_t}{1 - \mu_c K_t}(100)$$

or

$$\text{error} = \frac{K_t(\epsilon_t/\epsilon_a + \mu_c)}{1 - \mu_c K_t}(100) \qquad [\text{Eq. (2.15)}]$$

EXAMPLE 2.4

Determine the error introduced by neglecting the transverse sensitivity when a WK-06-500AF-350 strain gage is used to measure the axial strain in a tensile specimen having a Poisson's ratio of 0.33, if the Poisson's ratio of the caliberation beam material was 0.285.

Solution   From Table 2.2, the transverse sensitivity factor for a WK-06-500AF-350 strain gage is −8.0%. For a uniaxial tensile specimen, $\epsilon_t/\epsilon_a = -\mu = -0.33$. Using Eq. (2.15) yields

$$\begin{aligned}
\text{error} &= \frac{K_t(\epsilon_t/\epsilon_a + \mu_c)}{1 - \mu_c K_t} = \frac{K_t(-\mu + \mu_c)}{1 - \mu_c K_t} \\
&= \frac{(-0.08)(-0.33 + 0.285)}{1 - (0.285)(-0.08)} \\
&= \frac{0.0036}{1.0228} = 0.0035 = 0.35\%
\end{aligned}$$

Equation (2.16) is useful for correcting the transverse sensitivity effects. Corrections should always be made when strain gages with large transverse sensitivity factors are used or the measurements are done under biaxial stress fields. To correct for the transverse sensitivity effect, strains in two perpendicular directions are measured experimentally when the biaxial strain field is unknown. Let ϵ_a' and ϵ_t' be the indicated strains in the axial and transverse directions of the specimen. Using Eq. (2.16) we obtain

$$\begin{aligned}
\epsilon_a'(1 - \mu_c K_t) &= \epsilon_a + K_t \epsilon_t \\
\epsilon_t'(1 - \mu_c K_t) &= \epsilon_t + K_t \epsilon_a
\end{aligned} \qquad (2.17)$$

where ϵ_a and ϵ_t are the corrected (or actual) strains in the axial and transverse directions, respectively. Solving Eq. (2.17) for the corrected strains yields

$$\begin{aligned}
\epsilon_a &= \frac{1 - \mu_c K_t}{1 - K_t^2}(\epsilon_a' - K_t \epsilon_t') \\
\epsilon_t &= \frac{1 - \mu_c K_t}{1 - K_t^2}(\epsilon_t' - K_t \epsilon_a')
\end{aligned} \qquad (2.18)$$

Once the indicated strains ϵ_a' and ϵ_t' are known, the corrected strains ϵ_a and ϵ_t can be obtained using Eq. (2.18).

EXAMPLE 2.5

An aluminum 2024-T351 specimen is subjected to biaxial loading. At a certain load the strains indicated in the x and y directions are $5000\mu\epsilon$ and $200\mu\epsilon$, respectively. The type of strain gage is KFE-2-C1, which is made by Kyowa Electronic Instruments Company (Japan). The values $\mu_c =$

0.290 and $K_t = 0.4\%$ are given by the manufacturer. Calculate the corrected strains and percentage errors in the x and y directions of the specimen.

Solution   Since $\mu_c = 0.290$, $K_t = 0.004$, $\epsilon'_a = 5000\mu\epsilon$, and $\epsilon'_t = 200\mu\epsilon$ for a gage mounted in the axial direction of the specimen, substitution into Eq. (2.18) yields

$$\epsilon_x = \frac{1 - (0.290)(0.004)}{1 - (0.004)^2}[5000 - (0.004)(200)] = 4999.6\mu\epsilon$$

Using Eq. (2.15), the percentage error involved in neglecting the transverse sensitivity of the gage mounted in the x direction of the specimen is given by

$$error = \frac{5000 - 4999.6}{4999.6}(100) = -0.008\%$$

It is clear that the error is negligible since ϵ_y is small. Similarly, $\epsilon'_a = 200\ \mu\epsilon$ and $\epsilon'_t = 5000\mu\epsilon$ for the gage mounted in the y direction of the specimen. Substitution into Eq. (2.18) gives

$$\epsilon_y = \frac{1 - (0.290)(0.004)}{1 - (0.004)^2}[200 - (0.004)(5000)] = 180.0\mu\epsilon$$

If transverse sensitivity is neglected, the percentage error of the gage in the y direction is

$$error = \frac{200 - 180}{180}(100) = 11.1\%$$

Thus the error is 11.1% if transverse sensitivity is not considered. The error is larger in the gage mounted in the y direction of the specimen than in the gage mounted in the x direction of the specimen, since the ratio ϵ_t/ϵ_a $(= \epsilon_x/\epsilon_y)$ is much larger for the gage in the y direction (approximately 25).

Equation (2.15) indicates that the error is a function of both K_t and the ratio ϵ_t/ϵ_a, as shown graphically in Fig. 2.4 (for $K_t > 0$). The error can be significant when K_t and/or ϵ_t/ϵ_a are large, as shown in Examples 2.4 and 2.5, so it is always important to account for the transverse sensitivity of the gage.

A calibration method for the transverse sensitivity factor K_t is also given in ASTM Standard E251 and described briefly here. Nearly identical gages are mounted on a specially designed calibrated beam, which is in a uniaxial strain field, parallel and perpendicular to the direction of the uniaxial strain. By measuring the resistance changes of both gages, K_t is determined as follows:

$$K_t = \frac{\Delta R_T/R_T}{\Delta R_L/R_L}(100) \tag{2.19}$$

where ΔR_T and ΔR_L are the resistance changes of gages, mounted in the transverse and longitudinal directions with initial resistances R_T and R_L, respectively. K_t is usually given as a percentage.

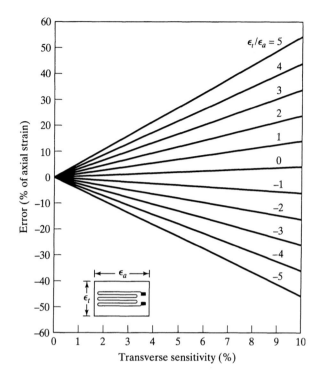

FIGURE 2.4   Percentage error in axial strain due to transverse sensitivity [5].

EXAMPLE 2.6

The transverse sensitivity factor K_t can also be determined by a simple tension test with identical gages mounted in the axial and transverse directions, as shown in Fig. E2.6. Derive an expression for K_t in terms of Poisson's ratio μ of the specimen material and the readings (ϵ_1 and ϵ_2) from the two gages.

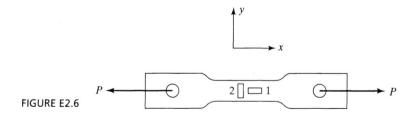

FIGURE E2.6

Solution   For simple tension,

$$\epsilon_x = \frac{\sigma}{E} \qquad \epsilon_y = \frac{-\mu\sigma}{E}$$

where E is Young's modulus of the specimen material. From Eq. (2.11) we have

$$\frac{\Delta R_1}{R_1} = S_g \epsilon_x + S_t \epsilon_y = \frac{\sigma}{E}(S_g - \mu S_t)$$

$$\frac{\Delta R_2}{R_2} = S_g \epsilon_y + S_t \epsilon_x = \frac{\sigma}{E}(S_g - \mu S_g)$$

where the subscripts 1 and 2 are associated with gages 1 and 2, as shown in Fig. E2.6, and S_a ($= S_{gl}$) and S_t ($= S_{gt}$) are gage sensitivities to the axial and transverse strain, respectively. Solving the foregoing equations simultaneously yields

$$S_t = \frac{E}{\sigma(1 - \mu^2)}\left(\frac{\Delta R_2}{R_2} + \mu \frac{\Delta R_1}{R_1}\right)$$

$$S_a = \frac{E}{\sigma(1 - \mu^2)}\left(\frac{\Delta R_1}{R_1} + \mu \frac{\Delta R_2}{R_2}\right)$$

$$K_t = \frac{S_t}{S_a} = \frac{(\Delta R_2/R_2) + \mu(\Delta R_1/R_1)}{(\Delta R_1/R_1) + \mu(\Delta R_2/R_2)}$$

Note that $\Delta R_1/R_1 = G_f \epsilon_1$ and $\Delta R_2/R_2 = G_f \epsilon_2$; thus

$$K_t = \frac{\epsilon_2/\epsilon_1 + \mu}{1 + \mu\epsilon_2/\epsilon_1}$$

EXAMPLE 2.7

A value of $K_t = 1\%$ was obtained by the method described in Example 2.6 with an assumed Poisson's ratio of 0.330 for the specimen material. Later it was found that the true Poisson's ratio was 0.325. What was the error introduced in the K_t?

Solution
From Example 2.6

$$K_t = \frac{\epsilon_2/\epsilon_1 + \mu}{1 + \mu\epsilon_2/\epsilon_1}$$

Thus

$$\frac{\epsilon_2}{\epsilon_1} = \frac{\mu - K_t}{K_t \mu - 1} = \frac{0.330 - 0.01}{(0.01)(0.330) - 1} = -0.32106$$

For $\mu = 0.325$,

$$K_t = \frac{-0.32106 + 0.325}{1 - (0.32106)(0.325)} = 0.0044$$

$$\text{error} = \frac{0.01 - 0.0044}{0.0044} = 1.27 = 127\%$$

It is very important to determine Poisson's ratio as accurately as possible whenever the method described in Example 2.6 is used to determine the transverse sensitivity factor K_t.

2.6   ENVIRONMENTAL EFFECTS ON METAL-FOIL STRAIN GAGES

The widely used metal-foil electrical resistance strain gages are small precision resistors mounted on a flexible carrier that can be bonded to a specimen. For a Kyowa (KFD-2-C1-23) strain gage, for example, the manufacturer's quoted gage resistance is accurate to ±0.25% and the gage factor is accurate to ±1.0%. These specifications indicate that precise strain measurements can obviously be achieved by using metal-foil electrical resistance strain gages. In practice, however, gage performance is dependent on such factors as the quality of the gage installation, the strain field to be measured, and the environmental conditions during the experiment. One must be aware of these effects and take them into account when installing a gage and analyzing test data. Some of the factors frequently encountered that affect results directly are discussed below.

Temperature Changes

Temperature fluctuations will cause changes in gage resistance. Because of this purely temperature-induced resistance change, a strain, called the *apparent strain* to distinguish it from strain in the specimen due to the applied load, will be registered by the strain indicator. The apparent strain caused by the temperature change is perhaps the most serious and prominent source of error in static strain measurements with strain gages and deserves much more careful consideration. Two effects in the strain gage will cause the apparent strain when the ambient temperature changes: (1) the effect due to the difference in thermal expansion coefficients between the gage grid and the base material, and (2) the effect due to the temperature dependence of the electrical resistivity of the grid conductor [7]. After correcting for the transverse sensitivity effect, the apparent strain ϵ_{app} can be expressed as [7]

$$\epsilon_{app} = \left[\frac{1 + K_t(\alpha_m - \alpha_g)}{(1 - \mu_c k_t)} + \frac{\beta}{G_f} \right] \Delta T \qquad (2.20)$$

where β is the temperature coefficient of resistivity of the gage grid, G_f is the gage factor, α_m is thermal expansion coefficient of the base material, α_g is thermal expansion coefficient of the gage grid, ΔT is temperature change from an arbitrary initial reference temperature, K_t the transverse sensitivity factor of the gage, and μ_c is Poisson's ratio of the calibration beam.

Whenever the thermal expansion coefficient of the grid differs from that of the base material (i.e., $\alpha_g \neq \alpha_m$) the gage is subjected to a mechanical strain $\epsilon = (\alpha_m - \alpha_g) \Delta T$ when there is a temperature change ΔT [Eq. (2.20)]. This occurs because the base material expands or contracts and the gage, bonded on it firmly, is forced to undergo the same expansion or contraction. Only when the gage grid and the base material have identical thermal expansion coefficients will this temperature-induced mechanical strain be zero. However, the gage may still register an apparent strain if the temperature coefficient of resistivity β is not zero.

Unfortunately, it is impossible to separate the apparent strain caused by the temperature change from the strain caused by the applied load. However, there are two methods that can be employed in actual practice to eliminate completely the apparent

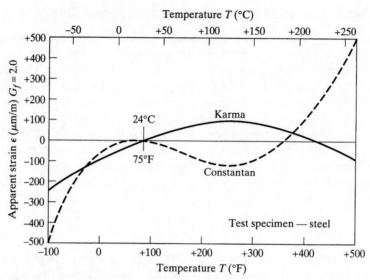

FIGURE 2.5   Apparent strain–temperature curves [7].

strain induced by the temperature change. The first way is to select a material for the gage conductor so that the net quantity $[\{(1 + K_t)/(1 - \mu_c K_t)\}(\alpha_m - \alpha_g) + \beta/G_f]$ is zero [i.e., both terms in Eq. (2.20) are zero or they cancel each other out]. This will produce *self-temperature-compensated strain gages,* which are commercially available. The second way is to compensate for the effects of the temperature change in the Wheatstone bridge or potentiometer circuit used to measure the output of the gage. This method is discussed in detail in Chapter 3.

It should be mentioned that temperature-compensated gages are not perfectly compensated for a wide range of temperatures, due to the nonlinear behavior of both the thermal expansion coefficient and the temperature coefficient of resistivity. Figure 2.5 shows the apparent strain–temperature curves for karma and constantan alloys. Obviously, the apparent strain is small only when the temperature change is within the neighborhood of 24C° (75°F).

EXAMPLE 2.8

Two Micro-Measurements self-temperature-compensated gages, MA-06-062TT-120, were mounted on an aluminum tube in the axial and circumferential directions. At room temperature (23.4°C), both strain readings are zero. When the temperature was increased to 65.3°C, the measured apparent strains are $500\mu\epsilon$ and $501\mu\epsilon$ in the axial and circumferential directions, respectively. Compare the experimental data with the theoretical values. The thermal expansion coefficient of aluminum is 23.2 ppm/°C (assuming for simplicity that $K_t = 0$).

Solution   According to designations for the Micro-Measurements self-temperature-compensated gage, the two-digit S-T-C number (e.g., 06) means that the thermal expansion coefficient of the material is 6 ppm/°F [7]. In the vicinity of room temperature, for material with $\alpha_m = 6 \times 10^{-6}$, $\epsilon_{app} = [(\alpha_m - \alpha_g) + \beta/G_f] \Delta T \approx 0$. Thus $\alpha_g - \beta/G_f = 6$ ppm/°F $= 10.8$ ppm/°C. Due to the self-temperature-compensating mismatch, the theoretical apparent strain is given by

$$\epsilon_{app} = (23.2 - 10.8)(65.3 - 23.4) = 520 \mu\epsilon$$

$$error = \frac{500 - 520}{500} = -4.0\%$$

It is clear that the error is caused by the nonlinear characteristic of both expansion and resistivity coefficients for temperature, as shown in Fig. 2.5.

EXAMPLE 2.9

Derive a general expression for apparent strain ϵ_{app} due to a S-T-C mismatch in terms of the S-T-C number n of the strain gage, the thermal expansion coefficient α_m of the base material, the temperature change ΔT, and the apparent strain $\bar{\epsilon}_{app}$, which is the strain when there is no S-T-C mismatch and can be obtained from Fig. 2.5 at temperature T. Note that $\bar{\epsilon}_{app} = 0$ at approximately room temperature, which is the case in Example 2.8 ($K_t = 0$).

Solution   By Eq. (2.20),

$$\bar{\epsilon}_{app} = \left[(\bar{\alpha}_m - \alpha_g) + \frac{\beta}{G_f} \right] \Delta T \approx \left[(n - \alpha_g) + \frac{\beta}{G_f} \right] \Delta T \qquad (E1)$$

$$\epsilon_{app} = \left[(\alpha_m - \alpha_g) + \frac{\beta}{G_f} \right] \Delta T \qquad (E2)$$

From Eq. (E1) we have

$$\left(-\alpha_g + \frac{\beta}{G_f} \right) \Delta T = \bar{\epsilon}_{app} - n \Delta T \qquad (E3)$$

Substituting Eq. (E3) into Eq. (E2) gives

$$\epsilon_{app} = \bar{\epsilon}_{app} + (\alpha_m - n) \Delta T \qquad (E4)$$

Zero Shift in Strain Cycling

When the strain measured by the strain gage is plotted against the applied strain in a strain cycle, a slight deviation from linearity is usually observed and the unloading curve is not coincident with the loading curve, so a hysteresis loop is formed. The curve is similar to a stress–strain curve for an elastoplastic material. The negative strain, indicated when the applied strain is returned to zero, is called *zero shift*. Cumulative zero shift is one of the factors that must be considered when a strain gage is used in a fatigue or cyclic loading experiment. The magnitude of the zero shift, hysteresis, and deviation from linearity depend on the degree of plastic deformation in the foil material, the carrier material, the strain level, and the quality of the bond. For a properly installed gage system, the magnitude is usually small and requires no correction.

Moisture and Humidity Effects

A strain gage installation can be affected detrimentally in several ways by direct exposure to moisture. For example, the gage-to-ground resistance will be decreased and the strength and rigidity of the adhesive bond will be degraded. Unprotected gages may be used briefly under favorable conditions of humidity and cleanliness or when restricted to the laboratory. However, if it is to be used over extended periods, any strain gage installation must have some protection to exclude moisture.

Various methods for waterproofing strain gages have been developed and used successfully. The method adopted depends on the application and the extent of the gage exposure to the moisture. For a normal laboratory experiment, a thin layer of air-drying polyurethane coating (e.g., M-COAT A, supplied by Micro-Measurements, Inc.) is usually sufficient to protect the gage installation from moisture in the air for short-term use. For strain gage installations in severe environments, additional protection is required. Figure 2.6 shows an example of strain gage protection under an extremely high-pressure water environment. For really long-term strain gage installations in severe environments, the addition of other strain gage protection should be employed. A semihard cover (either metal or plastic) protects it from any mechanical damage and acts as a first-stage barrier against the ingress of moisture. A small amount of silica gel makes the trapped air (if any) as dry as possible [8]. Waterproofed gages (e.g., KFW series gages by Kyowa) and fully encapsulated gages (e.g., WA, WK, and WD series gages by Micro-Measurements) are commercially available.

Gage Factor Variation with Temperature

The gage factor changes with temperature fluctuations. The error due to this effect depends on the gage material involved and the test temperature. In some cases, the error introduced by gage factor variation is small enough to be neglected. In others, correction may be necessary. Figure 2.7 shows the variation of gage factor with temperature change for constantan and isoelastic alloys. From Fig. 2.7 it can be seen that the effect in the constantan alloy is small and essentially linear. The effect in the isoelastic alloy is essentially nonlinear and correction may be required, depending on the test temperature and other conditions (i.e., dynamic or static test).

In some instances, for convenience, the gage factor is deliberately set on the strain indicator to a value that differs from the value given by the manufacturer, thus

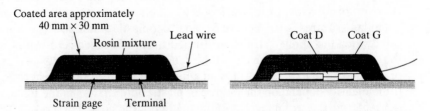

FIGURE 2.6 Strain gage protection arrangement for long-term installations in a severe environment [14].

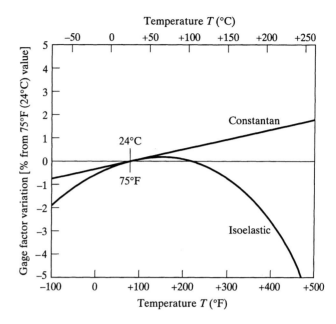

FIGURE 2.7   Gage factor variation with temperature [7].

making corrections necessary. In principle, any measured strain can be corrected from one gage factor to another by the equation

$$\epsilon = \epsilon_i \frac{G_{f1}}{G_f} \tag{2.21}$$

where ϵ is the corrected strain, ϵ_i the strain recorded from the strain indicator, G_{f1} the gage factor used in the strain indicator, and G_f the correct gage factor.

EXAMPLE 2.10

For convenience, the gage factor 2.00 is set on the strain indicator, since gages with different gage factors are employed during the experiment. At a certain load, one gage output is $2000\mu\epsilon$. The actual gage factor is 2.18. What is the corrected strain?

Solution   Let ϵ be the corrected strain. By using Eq. (2.21), we obtain

$$\epsilon = (2000.0)\left(\frac{2.0}{2.18}\right) = 1834.9\mu\epsilon$$

Therefore, the corrected strain is $1835.0\mu\epsilon$.

There are several other factors [17] that can affect gage performance, and they should be considered. If the proper value of excitation voltage is considerably exceeded, the gage performance will be degraded. When strain is measured at cryogenic temperatures, gage factor variation with temperature, as well as temperature-induced apparent strain, must be accounted for. While in high-temperature strain measurement, the resistance of a strain gage R is a function of temperature T, time t, and strain ϵ, namely, $R = f(T,t,\epsilon)$. In fatigue or strain cycling tests, changes in the gage factor, failure of the gage in fatigue, and the zero shift are major factors. In the case of hydrostatic high-pressure applications, for instance in the stress analysis of pressure vessels and piping systems, pressure-induced resistance changes must be taken into account in analysis of the measured strain data, or self-compensation methods, which compensate for temperature and pressure effects simultaneously, should be used [14]. When large strains are measured by bonded strain gages, it is very important to correct for errors due to Wheatstone bridge nonlinearity [15] and gage factor variation with strains [16], discussed in Chapter 3.

PROBLEMS

2.1. For very large strains, a special liquid-metal strain gage can be employed, which is simply a rubber tube filled with liquid metal [3]. It is obvious that the volume of the liquid metal, $V = LA$, will not change during deformation of the test specimen. Assuming that the change in the specific resistance of the liquid metal is negligible, show that $\Delta R/R = 2\epsilon + \epsilon^2$, where $\epsilon = \Delta L/L$.

2.2. When the strains are larger than 1%, deformation of a constantan alloy becomes elastic–plastic. However, experiment shows that the relationship between $\Delta R/R$ and ϵ is linear up to 8% (or more) of strains. Discuss the reasons for this remarkable linearity. Poisson's ratio will increase from 0.3 to 0.5 as deformation of the alloy changes from elastic to fully plastic. (*Hint:* Consider the changes in $\Delta r/r$.)

2.3. In general, the stress gage is designed only for a particular Poisson's ratio. If Poisson's ratio of the material is different from the specified value of the stress gage, the simplest way to measure the stress directly is to bond a conventional gage at a specific angle ϕ to the axis of principal stress σ_1, as shown in Fig. P2.3. Determine **(a)** the angle ϕ in terms of μ, and **(b)** $\Delta R/R$ in terms of S_a, E, μ, and σ_1, where E and μ are Young's modulus and Poisson's ratio of the specimen, respectively, S_a the strain sensitivity, and R the resistance of the strain gage. Assume that the $\Delta R/R$ value of a gage is the same as that of a single conductor.

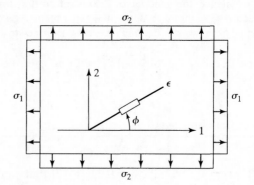

FIGURE P2.3

2.4. A 350-Ω electrical resistance strain gage is mounted on a cantilever beam, as shown in Fig. P2.4. The beam is subjected to a concentrated load P at its free end. At $P = 130$ lb, the gage undergoes a change in resistance of 0.275 Ω. What is the gage factor? Assume that $L_1 = 8$ in., $b = 0.4$ in., $h = 2.0$ in., and Young's modulus of the beam material is 10,000 ksi.

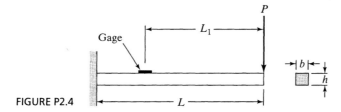

FIGURE P2.4

2.5. A 120-Ω strain gage is mounted on a tensile specimen with the gage axis 20° from the axial direction of the specimen, as shown in Fig. P2.5. The gage factor is 2.05. What will be the resistance change of the gage if the specimen is subjected to a tensile load of 1000 lb? Young's modulus and Poisson's ratio of the specimen are 10,000 ksi and 0.33, respectively.

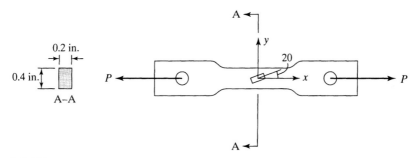

FIGURE P2.5

2.6. In a uniaxial tension test, a strain gage with 120 Ω is mounted on the specimen along the axial direction to record the axial strain. At a certain load, the reading of the strain gage is 9970$\mu\epsilon$ and the corresponding gage resistance is 122.53 Ω. What is the gage factor?

2.7. A circular-arc gage [10] with resistance R (shown in Fig. P2.7) is mounted on the surface of a specimen in the biaxial strain state. Determine $\Delta R/R$ in terms of ϵ_x, ϵ_y, and γ_{xy}. The radius of the circular arc is r.

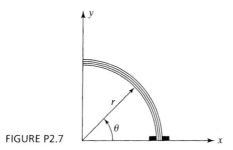

FIGURE P2.7

2.8. Determine the error introduced by neglecting the transverse sensitivity if a WK-06-500AF-350 strain gage is used to measure the transverse strain in a tensile specimen having a Poisson ratio of 0.33.

2.9. An aluminum 6061 specimen, shown in Fig. P2.9, is subjected to a tensile loading. Two KFE-2-C1 gages were mounted on the specimen in the axial and transverse directions. At load $P = 1109$ lb, the strain readings are $2265\mu\epsilon$ and $-703\mu\epsilon$ for gages 1 and 2, respectively. Calculate the corrected strains and percentage errors in the axial and transverse directions of the specimen if $K_t = 0.4\%$ and $\mu_c = 0.290$. What is Poisson's ratio for aluminum 6061?

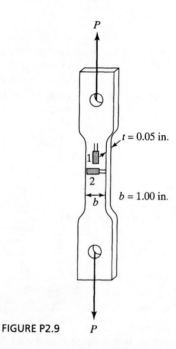

FIGURE P2.9 P

2.10. An MA-06-062TT-120 gage is made of constantan alloy. What should be the error in Example 2.8 if Eq. (E4) is employed? The gage factor was set at 2.0 on the strain indicator.

2.11. Two gages taken from the same package were mounted on an aluminum tube and a steel bolt, respectively. When the temperature changes from 75°F to 150°F, the apparent strain is $230\mu\epsilon$ from the gage mounted on the tube. What will be the apparent strain registered by the gage mounted on the bolt when the bolt undergoes the same temperature change? The thermal expansion coefficients for aluminum and steel are 12.9 and 6.0 ppm/°F, respectively.

2.12. Strain gages 1 and 2 were bounded on the surface of a cylindrical pressure vessel along the axial and circumferential directions. The specifications are as follows:

Gage number	Gage factor	$R\,(\Omega)$	$K_t\,(\%)$
1	2.09	120	1.3
2	2.06	120.3	1.3

For convenience, the gage factor was set at 2.00 on the strain indicator. At a pressure of 500 psi, the readings are 242$\mu\epsilon$ and 128$\mu\epsilon$ for gages 1 and 2, respectively. What are the corrected strains?

REFERENCES

[1] C. C. Perry, The resistance strain gage revisited, W. Murray Lecture, *Proc. 5th Int. Congr. Exp. Mech.*, Montreal, Quebec, Canada, June 10–15, 1984.

[2] W. Thomson (Lord Kelvin), On the electrodynamic qualities of metals, *Proc. R. Soc.*, 1856.

[3] J. W. Dally and W. F. Riley, *Experimental Stress Analysis*, 2nd ed., McGraw-Hill, New York, 1978.

[4] W. M. Murray and P. K. Stein, *Strain Gage Techniques*, 1958.

[5] *Transverse Sensitivity Errors*, Tech Note TN-509, Measurements Group, Inc., Raleigh (North Carolina), 1982.

[6] A. S. Kobayashi, ed., *Handbook on Experimental Mechanics*, Second Edition, Society for Experimental Mechanics, Bethel, CT, 1993.

[7] *Strain Gage Temperature Effects*, Tech. Note TN-504-1, Measurements Group, Inc., Raleigh (North Carolina), 1989.

[8] J. Vaughan, *Application of B & K Equipment to Strain Measurements*, printed in Denmark, 1975.

[9] *Bondable Terminals*, Tech Tip TT-603, Measurements Group, Inc., Raleigh (North Carolina), 1983.

[10] *Plane-Shear Measurement*, Tech Note TN-512, Measurements Group, Inc., 1983.

[11] G. Sines, *Elasticity and Strength*, Allyn & Bacon, Needham Heights, Mass., 1969.

[12] *Strain Gage and Temperature Sensor: Instruction Manual* (revised), Kyowa Electronic Instruments Company, Tokyo, 1982.

[13] *Strain Gage Selection*, Tech Note TN-502-2, Measurements Group, Inc., Raleigh (North Carolina), 1989.

[14] A. S. Khan and J. C. Chen, Further study on the use of foil strain gages under extremely high-pressure water environment, *Exp. Mech.*, **25**, pp. 123–128, 1985.

[15] *Wheatstone Bridge Nonlinearity*, Tech Note TN-507, Measurements Group, Inc., Raleigh (North Carolina), 1982.

[16] *High-Elongation Measurements*, Tech Tip TT-605, Measurements Group, Inc., Raleigh (North Carolina), 1983.

[17] *Catalog 500*, Part B, Precision Strain Gages, Measurements Group, Inc., Raleigh (North Carolina), 1988.

[18] *Strain Gage Installations with M-Bond 200 Adhesive*, Instr. Bull. B-127-11, Measurements Group, Inc., Raleigh (North Carolina), 1979.

[19] *Strain Gage Applications with M-Bond AE-10/15 and M-Bond GA-2 Adhesive Systems*, Instr. Bull. B-137-13, Measurements Group, Inc., Raleigh (North Carolina), 1979.

C H A P T E R   3

Strain Gage Circuitry, Transducers, and Data Analysis

3.1   INTRODUCTION

The electrical resistance strain gage is the most versatile of many devices to measure strains on the surfaces of machine components and structural members. Because resistance change in a strain gage is usually very small, it cannot be measured accurately with an ordinary ohmmeter. The Wheatstone bridge is widely used in practice; one or more of the four arms of the bridge are strain gages. In this chapter, the basic principles of the Wheatstone bridge and the potentiometer are described. Such effects as due to lead wire resistance are discussed briefly. Some transducers, in which the strain gages are used as the sensing elements, are considered. Strain gage data analyses are presented. Finally, the nonlinearity of Wheatstone bridge and the variations of gage factor are discussed briefly, because they are very important when large strains are to be measured by using strain gages.

3.2   WHEATSTONE BRIDGE

The Wheatstone bridge [1] is a basic circuit employed to measure extremely small resistance changes in a strain gage when it is subjected to a strain. Figure 3.1 shows a constant-voltage Wheatstone bridge that is normally used to record strain gage outputs in static and dynamic applications. Let us consider briefly the principle of operation of the Wheatstone bridge. For the circuit shown in Fig. 3.1, the voltage drops across R_1 and R_4, denoted by V_{ab} and V_{ad}, respectively, are given by the equations

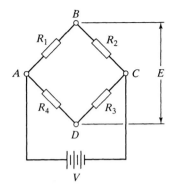

FIGURE 3.1 Constant-voltage Wheatstone bridge.

$$V_{ab} = \frac{R_1}{R_1 + R_2} V \qquad V_{ad} = \frac{R_4}{R_3 + R_4} V$$

where V is the applied voltage across the bridge.

The voltage output of the bridge E is represented by

$$E = V_{ab} - V_{ad} = \frac{R_1 R_3 - R_2 R_4}{(R_1 + R_2)(R_3 + R_4)} V \qquad (3.1)$$

It is clear that the output voltage of the bridge is zero (i.e., the bridge is *balanced*) when the term $R_1 R_3 - R_2 R_4$ is zero or when

$$R_1 R_3 = R_2 R_4 \qquad (3.2)$$

Equation (3.2) represents a very important relationship which indicates that any change in the resistance in one arm of the bridge can be balanced by adjusting the resistance(s) in the other arms of the bridge. Figure 3.2 shows some of the typical balancing arrangements.

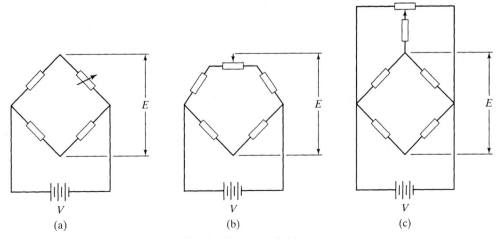

(a)　　　　　　　(b)　　　　　　　(c)

FIGURE 3.2 Typical methods of balancing the Wheatstone bridge.

Consider an initially balanced bridge, namely, $R_1 R_3 = R_2 R_4$, so that $E = 0$, and then resistances R_1, R_2, R_3, and R_4 are changed by the amounts $\Delta R_1, \Delta R_2, \Delta R_3$, and ΔR_4. The voltage output ΔE of the bridge can be determined using Eq. (3.1); that is,

$$\Delta E = \frac{(R_1 + \Delta R_1)(R_3 + \Delta R_3) - (R_2 + \Delta R_2)(R_4 + R_4)}{(R_1 + \Delta R_1 + R_2 + \Delta R_2)(R_3 + \Delta R_3 + R_4 + \Delta R_4)} V$$

Using Eq. (3.2), neglecting second-order terms (e.g., $\Delta R_1 \, \Delta R_3$) and relatively smaller terms (e.g., $R_1 \Delta R_3$) in the denominator, it can be shown that

$$\Delta E = V \frac{R_1 R_2}{(R_1 + R_2)^2} \left(\frac{\Delta R_1}{R_1} - \frac{\Delta R_2}{R_2} + \frac{\Delta R_3}{R_3} - \frac{\Delta R_4}{R_4} \right) \tag{3.3}$$

Let $R_2/R_1 = m$; then Eq. (3.3) can be rewritten as

$$\Delta E = V \frac{m}{(1 + m)^2} \left(\frac{\Delta R_1}{R_1} - \frac{\Delta R_2}{R_2} + \frac{\Delta R_3}{R_3} - \frac{\Delta R_4}{R_4} \right) \tag{3.4}$$

Equation (3.4) is the basic equation governing the strain measurement of a Wheatstone bridge. Note that the second-order terms should be included when the strains being measured are greater than 5%. The correction for nonlinearity of the Wheatstone bridge will be discussed in Section 3.6. The sensitivity of the Wheatstone bridge, S_w, can be defined as,

$$S_w = \frac{\Delta E}{\epsilon} = \frac{V}{\epsilon} \frac{m}{(1 + m)^2} \left(\frac{\Delta R_1}{R_1} - \frac{\Delta R_2}{R_2} + \frac{\Delta R_3}{R_3} - \frac{\Delta R_4}{R_4} \right) \tag{3.5}$$

Normally, a fixed voltage is applied to the bridge, then the sensitivity of the bridge depends on the number of active arms employed, the gage factor G_f, and the resistance ratio m. However, a voltage magnitude can be selected to increase circuit sensitivity. The upper limit of the voltage is determined by the power dissipated by the strain gage(s) used in the bridge. Once a particular gage type is selected, the gage factor cannot be varied to increase sensitivity. Finally, the two most important parameters are the number of active arms and the resistance ratio. Quarter-, half-, and full-bridge arrangements are obtained if one, two, and four active strain gage arms are employed, respectively; these are discussed in some detail below.

Quarter Bridge

The arrangement is called a *quarter bridge* when only one active strain gage is used, as shown in Fig. 3.3. In this figure, R_g is the active strain gage, which undergoes the same deformation as the structure, and R_d is the dummy or inactive gage, which is identical to the active gage but does not encounter any mechanical strains and is used for compensating the temperature effect; the other two arms contain fixed resistors. Any resistance change ΔR_g in the gage will disturb the balance of the bridge. The change of resistance in the active gage may be produced by mechanical loading and/or by temperature changes if the strain gage material's thermal expansion coefficient differs from that of the structure's material undergoing strain analyses. If the bridge is initially balanced, from Eq. (3.4) the out-of-balance voltage ΔE will be

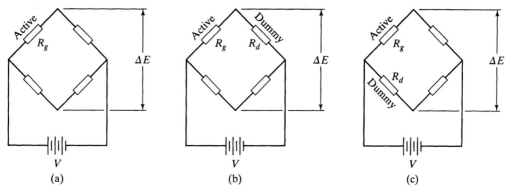

FIGURE 3.3 Quarter-bridge arrangements.

$$\Delta E = \frac{Vm}{(1+m)^2} \frac{\Delta R_g}{R_g} = \frac{V\epsilon G_f m}{(1+m)^2} \tag{3.6}$$

or

$$\epsilon = \frac{(1+m)^2 \, \Delta E}{VmG_f} \tag{3.7}$$

Equation (3.7) shows that it is only necessary to measure the out-of-balance voltage to obtain the strain when the bridge excitation voltage and the gage factor are known.

As mentioned earlier, temperature change will cause gage-resistance change so that apparent strain will be introduced if the thermal expansion coefficients of gage and structure or machine element materials differ. The simplest and most commonly employed method to eliminate this problem is to use a dummy gage in the Wheatstone bridge, as shown in Fig. 3.3b or c. The dummy strain gage is of the same type as the active gage but is bonded to an unstressed part of a structure or machine element, or on a separate piece of the same material as the structure or machine element under stress analysis; the active and dummy gages are located in close proximity so that they experience the same temperature change. It is easy to show that the temperature-induced apparent strain will be canceled out, and the bridge out-of-balance voltage is related only to the strain due to mechanical loads.

Using Eqs. (3.5) and (3.6), we can write the sensitivity of the fixed-voltage quarter bridge as

$$S_{wq} = \frac{m}{(1+m)^2} VG_f \tag{3.8}$$

Sometimes a variable voltage, whose upper limit is determined by the power dissipated by the strain gage(s) placed in the bridge, is applied to the bridge to increase the circuit sensitivity. For a variable-voltage quarter bridge, the voltage is related to the power dissipation by a single gage in one arm of the bridge as follows:

$$V = I_g(R_g + R_2) = I_g R_g(1 + m) = (1 + m)(P_g R_g)^{1/2}$$

where I_g is the current in the active gage and P_g is the power dissipation of the gage, respectively. Therefore, Eq. (3.8) can be rewritten as

$$S_{wq} = \frac{m}{1 + m} G_f (P_g R_g)^{1/2} \tag{3.9}$$

It is obvious that the sensitivity of the bridge is affected by two factors: the gage selection controlling $G_f(P_g R_g)^{1/2}$ and the circuit efficiency, denoted by $m/(1 + m)$. It is easy to show that the circuit efficiency for arrangements in Fig. 3.3a and c is the same and can be increased by selecting a high value for m (e.g., 80% with $m = 4$ and 90% with $m = 9$). The supply voltage remains within reasonable limits, while the circuit efficiency of the bridge, shown in Fig. 3.3b, is fixed and equal to 50%, since $m = 1$ due to the location of the dummy gage on arm 2. Thus the arrangement in Fig. 3.3c for temperature compensation is better than that in Fig. 3.3b because the circuit efficiency remains the same as that in Fig. 3.3a.

EXAMPLE 3.1

A strain gage is bonded to a tensile specimen along the axial direction to measure strain in that direction. To eliminate the effect of temperature change, a compensating gage (or dummy) is mounted on a separate piece of the same material as the specimen and placed close to it, as shown in Fig. E3.1. Prove that the apparent strains due to temperature change ΔT will be canceled out if a quarter bridge, also shown in Fig. E3.1, is used.

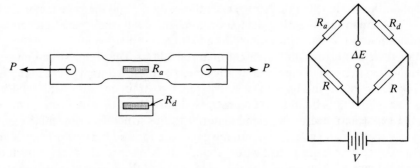

FIGURE E3.1

Solution   During the application of a tensile load P and temperature change ΔT, strain gage 1 will undergo a resistance change consisting of two parts, namely,

$$\frac{\Delta R_1}{R_1} = \frac{\Delta R_{aP} + \Delta R_{aT}}{R_a}$$

where ΔR_{aP} and ΔR_{aT} represent resistance change of the active gage due to load P and temperature change ΔT, respectively. However, strain gage 2 will undergo a resistance change due solely to temperature change. Thus

$$\frac{\Delta R_2}{R_2} = \frac{\Delta R_{dT}}{R_d}$$

By using Eq.(3.4) and since $\Delta R_3 = \Delta R_4 = 0$, we obtain

$$\Delta E = V \frac{m}{(1+m)^2} \left(\frac{\Delta R_1}{R_1} - \frac{\Delta R_2}{R_2} \right)$$

$$= V \frac{m}{(1+m)^2} \left(\frac{\Delta R_{ap}}{R_a} + \frac{\Delta R_{aT}}{R_a} - \frac{\Delta R_{dT}}{R_d} \right)$$

Obviously, $\Delta R_{aT}/R_a = \Delta R_{dT}/R_d$ if two gages are identical ($R_a = R_d$), and they undergo identical temperature change ΔT (i.e., $\Delta R_{aT} = \Delta R_{dT}$). Thus

$$\Delta E = \frac{V}{4} \frac{\Delta R_{ap}}{R_a}$$

which is independent of temperature change ΔT. Temperature compensation can also be achieved by employing a half- or a full-bridge circuit, which will be illustrated later.

EXAMPLE 3.2

Two strain gages are placed in a series in arm R_1 of a fixed-voltage quarter Wheatstone bridge, shown in Fig. E3.2. Assume that two gages experience the same strains. Determine the circuit sensitivity and compare it with the value obtained if a single gage is employed.

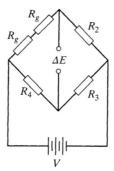

FIGURE E3.2

Solution For a fixed-voltage quarter Wheatstone bridge with a single active gage, the circuit sensitivity S_{wq} is given by Eq. (3.8), namely,

$$S_{wq} = \frac{m}{(1+m)^2} VG_f = \frac{m}{(1+m)^2} \frac{V}{\epsilon} \frac{\Delta R_1}{R_1}$$

For a single strain gage, $\Delta R_1/R_1 = \Delta R_g/R_g$, where R_g is the resistance of a strain gage (usually 120 Ω or 350 Ω). For the case of two gages connected in a series,

$$\frac{\Delta R_1}{R_1} = \frac{\Delta R_g + \Delta R_g}{R_g + R_g} = \frac{\Delta R_g}{R_g}$$

Thus there is no increase in circuit sensitivity with two gages connected in a series. The circuit sensitivity S_{wq} in each case is given by

$$S_{wq} = \frac{m}{(1+m)^2} \frac{V}{\epsilon} \frac{\Delta R_g}{R_g}$$

EXAMPLE 3.3

Suppose that the voltage can be varied in the Wheatstone bridge shown in Fig. E3.2. **(a)** Compare the circuit sensitivity for the two gages connected in a series with the one for a single gage. **(b)** Determine the percent change in circuit sensitivity for the two gages over a single gage. **(c)** Determine the required voltage for the two gages and for a single gage if $P_g = 0.03$ W, $R_g = 120\ \Omega$, and $m = 7$.

Solution   **(a)** For a variable-voltage quarter bridge, the sensitivity is given by Eq. (3.9), namely,

$$S_{wq} = \frac{m}{1 + m}\, G_f (P_1 R_1)^{1/2}$$

For a single active gage, $P_1 = P_g$ and $R_1 = R_g$; thus

$$S^1_{wq} = \frac{m}{1 + m}\, G_f (P_g R_g)^{1/2}$$

For two gages in a series, $P_1 = 2P_g$ and $R_1 = 2R_g$; therefore,

$$S^2_{wq} = \frac{m}{1 + m}\, G_f [(2P_g)(2R_g)]^{1/2} = 2\left(\frac{m}{1 + m}\right) G_f (P_g R_g)^{1/2} = 2S^1_{wq}$$

So the sensitivity for the two gages in a series is twice that for a single gage. It will be seen shortly that this is due to the use of a higher voltage; two gages permit the use of a voltage twice in magnitude for a single gage; the voltage magnitude is dictated by power dissipation in a single gage.

(b) Change % $= S^2_{wq} - S^1_{wq}/S^1_{wq}\ 100\% = 100\%$.

(c) $I_g = (P_g/R_g)^{1/2} = (0.03/120)^{1/2} = 0.0158$ A; for a single gage, $R_1 = R_g$,

$$V_1 = I_g(R_1 + R_2) = I_g R_g (1 + m) = (0.0158)\,(120)\,(1 + 7) = 15.18\ V$$

and for the two gages, $R_1 = R_g + R_g$,

$$V_2 = I_g R_1 (1 + m) = (0.0158)\,(120 + 120)\,(1 + 7) = 30.36\ V$$

Half Bridge

If the dummy gage in Fig. 3.3b or c is replaced by an active gage, as shown in Fig. 3.4, the resulting arrangement is called a *half bridge*. The half bridge has advantages for temperature compensation and higher bridge sensitivity over the quarter bridge so that small strain levels can be detected more accurately. As with the quarter bridge, the arrangement in Fig. 3.4b is superior to that in Fig. 3.4a, since the circuit efficiency in Fig. 3.4b is usually higher if a variable voltage is applied to the bridge.

Figure 3.5a shows an example of half-bridge application: A circular specimen is subject to a tensile load. Gages *a* and *t* are mounted along the axial and transverse directions, respectively. The corresponding bridge connection is shown in Fig. 3.5b (the half bridge). Assuming that identical strain gages are used, gage *a* has a resistance increase of ΔR at some load, and gage *t* will have a resistance decrease of $-\mu\, \Delta R$ due to Poisson's effect, where μ is Poisson's ratio.

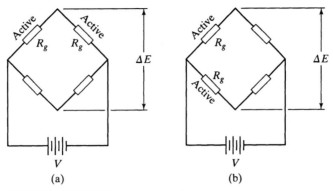

FIGURE 3.4 Half-bridge arrangements.

Using Eq. (3.4), the out-of-balance voltage ΔE becomes

$$\Delta E = V \frac{m}{(1+m)^2} \left(\frac{\Delta R_a}{R_a} - \frac{-\mu\, \Delta R_a}{R_a} \right) = \frac{(1+\mu)m}{(1+m)^2} VG_f\epsilon$$

Therefore, the sensitivity of the bridge is

$$S_{wh} = \frac{V(1+\mu)m}{(1+m)^2} G_f = \frac{(1+\mu)m}{1+m} G_f(P_gR_g)^{1/2} \qquad (3.10)$$

Comparing Eq. (3.10) with Eqs. (3.8) and (3.9), we find that the circuit efficiency increases $(1 + \mu)$ times. In practice, the principal advantage of this type of arrangement is to obtain automatic temperature compensation. Also, if two pairs of gages are mounted in a diametrically symmetric direction, then employing a half-bridge circuit by connecting the corresponding gages in series, or using a full bridge (as in tension–compression load cell), we can automatically cancel any bending effect. It should be mentioned that the self-temperature-compensating gages described earlier must be used when the specimen is large or when temperatures vary greatly.

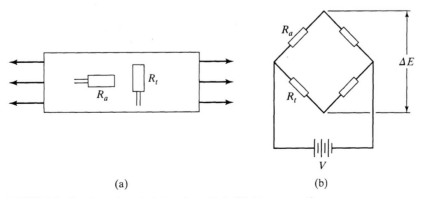

FIGURE 3.5 Specimen in uniaxial tension with half-bridge connection.

EXAMPLE 3.4

An assembly of a steel bolt and an aluminum tube is subjected to a temperature change ΔT. To measure the mechanical strain in the tube due to temperature increase ΔT, four gages are bonded in the longitudinal and circumferential directions, as shown in Fig. E3.4a; gages C and D are placed diametrically opposite to gages A and B, respectively. A half bridge with fixed voltage V is adopted. How should the four gages in the bridge be placed to eliminate the apparent strain due to temperature increase and to measure the mechanical strain? What is the output ΔE in terms of V, G_f, μ, and ϵ? (ϵ is the mechanical strain in the axial direction due to temperature increase ΔT.)

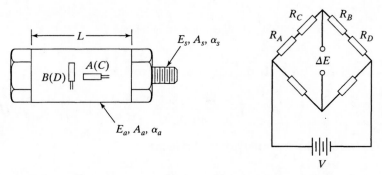

FIGURE E3.4

Solution   The four gages are positioned in the Wheatstone bridge as shown in Fig. E3.4b:

$$R_A = R_B = R_C = R_D = R_g$$

Since all gages experience the same temperature change, the resistance change ΔR_T due to a mismatch of the S-T-C number (thermal expansion coefficient) will be the same. Using Eq. (3.4), these resistance changes due to temperature will be canceled. In other words, the bridge is temperature compensated.

The mechanical strain in the axial direction is ϵ; it is $-\mu\epsilon$ in the circumferential direction due to Poisson's effect. Thus

$$\frac{\Delta R_1}{R_1} = \frac{\Delta R_A + \Delta R_C}{R_A + R_C} = \frac{2\Delta R_A}{2R_A} = G_f\epsilon$$

$$\frac{\Delta R_2}{R_2} = \frac{\Delta R_B + \Delta R_D}{R_B + R_D} = \frac{2\Delta R_B}{2R_B} = \frac{\Delta R_B}{R_B} = -G_f\mu\epsilon$$

Note that $m = 1$ and $\Delta R_3 = \Delta R_4 = 0$, and by using Eq. (3.4), $\Delta E = (V/4)G_f(1 + \mu)\epsilon$.

EXAMPLE 3.5

In Example 2.1 it was mentioned that the gage shown in Fig. E2.1 could also give readings proportional to shear strain γ if it was mounted along the axis of the shaft and connected in a half bridge, as shown in Fig. E3.5. Verify this statement.

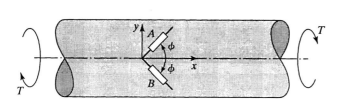

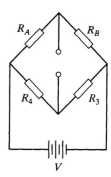

FIGURE E3.5

Solution   Using Eq. (1.34a), we have

$$\epsilon_A = \frac{\epsilon_x + \epsilon_y}{2} + \frac{\epsilon_x - \epsilon_y}{2}\cos 2\phi + \frac{\gamma}{2}\sin 2\phi$$

$$\epsilon_B = \frac{\epsilon_x + \epsilon_y}{2} + \frac{\epsilon_x - \epsilon_y}{2}\cos 2\phi - \frac{\gamma}{2}\sin 2\phi$$

$$\frac{\Delta R_A}{R_A} = G_f \epsilon_A$$

$$\frac{\Delta R_B}{R_B} = G_f \epsilon_B$$

Since $R_A = R_B = R$, $m = 1$. Note that $\Delta R_3 = \Delta R_4 = 0$, and by Eq. (3.4), we obtain

$$\Delta E = \frac{V}{4}\left(\frac{\Delta R_A}{R_A} - \frac{\Delta R_B}{R_B}\right) = \frac{VG_f}{4}\gamma \sin 2\phi$$

Full Bridge

The highest circuit efficiency can be achieved if four active gages are employed in the bridge, as shown in Fig. 3.6. Such an arrangement is called a *full bridge*. It is automatically temperature compensated when all four active gages bonded on the same mater-

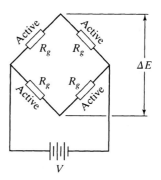

FIGURE 3.6   Full-bridge arrangement.

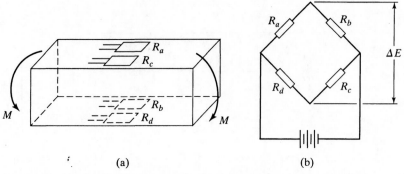

FIGURE 3.7 Specimen in bending with full-bridge arrangement.

ial, and are thus subjected to the same temperature changes and are of the same type. In this case also, bridge balancing should be obtained by using the arrangements shown in Fig. 3.2b or c. Use of a full bridge can increase the circuit efficiency up to 200% over that of a quarter bridge. The following example illustrates this.

Figure 3.7 shows four identical strain gages mounted on a beam with rectangular cross section in pure bending and with full-bridge connections. The strains at the top and bottom surfaces are the same in magnitude but opposite in sign. Therefore, gages a and c will have a resistance increase of ΔR during bending, while gages b and d will have a resistance decrease of $-\Delta R$. Since resistance changes due to the same temperature change will be canceled out automatically, by using Eq. (3.4) and noting that $m = 1$, the out-of-balance voltage ΔE becomes

$$\Delta E = \frac{V}{4}\left(\frac{\Delta R}{R} - \frac{-\Delta R}{R} + \frac{\Delta R}{R} - \frac{-\Delta R}{R}\right) = VG_f\epsilon$$

Therefore, the sensitivity of the bridge is

$$S_{wf} = VG_f = 2G_f(P_gP_g)^{1/2} \tag{3.11}$$

Note that this is possible only for the bending case where two gages measure positive strain and the other two determine an equal magnitude of negative strain.

If in addition to the pure bending moment, the structural member is subjected to a tensile force, the four gages will experience the same amount of resistance change due to this axial force, in addition to change caused by the bending moment. Using Eq. (3.4), it can be shown that the effect of the tensile load will be eliminated in the full-bridge arrangement, leaving only the bending components to be measured.

EXAMPLE 3.6

A load cell is to be designed to measure a tensile force P that is applied along the axis of the structural member (load cell), as shown in Fig. 3.7a. The bending moment, shown in this figure, may be produced due to slight misalignment of load with this axis. To measure the tensile strain caused by load P alone, how should the four gages be placed in a Wheatstone bridge? Neglect the effect of temperature change.

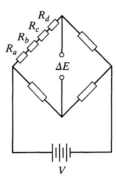

FIGURE E3.6

Solution   Since the temperature-change effect is negligible, a quarter bridge can be employed, as shown in Fig. E3.6 (all four gages are connected in a single arm). Let the resistance change caused by pure bending be ΔR_M and the resistance change caused by tensile load P be ΔR_P. Each gage has a resistance change ΔR, which is the sum of ΔR_P and ΔR_M.

For gages a and c,

$$\Delta R_i = \Delta R_P + \Delta R_M$$

and for gages b and d,

$$\Delta R_i = \Delta R_P - \Delta R_M$$

Therefore, by Eq. (3.4),

$$\Delta E = \frac{Vm}{(1+m)^2}\frac{\Delta R}{R} = \frac{Vm}{(1+m)^2}\frac{(\Delta R_P + \Delta R_M) + (\Delta R_P - \Delta R_M) + (\Delta R_P + \Delta R_M) + (\Delta R_P - \Delta R_M)}{R_a + R_b + R_c + R_d}$$

$$= \frac{Vm}{(1+m)^2}\frac{4\Delta R_P}{4R_g} = \frac{Vm}{(1+m)^2}\frac{\Delta R_P}{R_g} = \frac{Vm}{(1+m)^2}G_f\epsilon$$

where R_g, G_f, and ϵ are the gage resistance, the gage factor, and the axial strain, respectively, and $\Delta R/R$ represents the total resistance change per unit resistance in a single arm.

It should be pointed out here that there is no circuit sensitivity increase due to the use of four gages in a quarter bridge. Therefore, it is not economical to use four gages to measure strain in the axial direction. To eliminate any possible bending effect, only two gages are recommended in practice. These two gages should be connected in series and positioned in a quarter bridge. The best design for this load transducer to maximize the circuit sensitivity is two gages in the axial direction and two in the transverse direction with a full bridge.

EXAMPLE 3.7

Four strain gages were bonded on a cantilever beam and positioned in a fixed-voltage Wheatstone bridge as shown in Fig. E3.7. Prove that the bridge output is proportional to the applied bending moment M at the fixed end irrespective of the point of application of load P as long as it remains to the right of gage d.

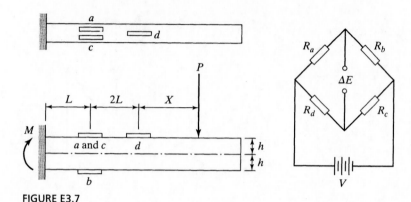

FIGURE E3.7

Solution   From Fig. E3.7 it can be seen that

$$\epsilon_a = -\epsilon_b = \epsilon_c = \frac{P(2L + X)h}{EI}$$

and

$$\epsilon_D = \frac{PXh}{EI}$$

where E is Young's modulus and I is the moment of inertia (or the second moment of area), respectively. Using Eqs. (3.4) and (2.9) and noting that $m = 1$, we obtain

$$\begin{aligned}
\Delta E &= \frac{V}{4}\left(\frac{\Delta R_a}{R_a} - \frac{\Delta R_b}{R_b} + \frac{\Delta R_c}{R_c} - \frac{\Delta R_d}{R_d}\right) \\
&= \frac{V}{4}(G_f\epsilon_a - G_f\epsilon_b + G_f\epsilon_c - G_f\epsilon_d) \\
&= \frac{G_f VPh}{4EI}[(2L + X) - (-2L - X) + (2L + X) - X] \\
&= \frac{G_f VPh}{2EI}(3L + X) \\
&= \frac{G_f Vh}{2EI}M = KM
\end{aligned}$$

where K is a constant and its definition is apparent.

3.3   CORRECTION FOR LONG LEAD WIRES

A Wheatstone bridge can be used to measure extremely small resistance changes in strain gages. Any small change in resistance of any one of the four arms in the bridge will affect the output voltage. If long lead wires have substantial resistance, their effect should be included in the circuit analysis for strain measurement [2]. Otherwise, errors could be introduced in such a measurement and signal attenuation and loss of temperature compensation could occur.

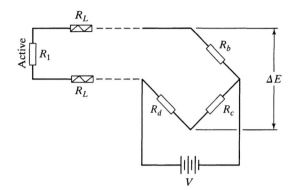

FIGURE 3.8 Two-lead-wire arrangement.

Consider first a single gage with two long lead wires, as shown in Fig. 3.8. From the figure it is clear that

$$R_1 = R_g + 2R_L$$

where R_g and R_L are gage and lead wire resistances, respectively. The lengths of the two lead wires are assumed to be the same, which is the usual case in practice. Note that

$$\frac{\Delta R_1}{R_1} = \frac{\Delta R_g}{R_g + 2R_L} = \frac{\Delta R_g/R_g}{1 + 2R_L/R_g} = \frac{G_f \epsilon}{1 + 2R_L/R_g} = G_f \epsilon' \qquad (3.12)$$

where G_f is the gage factor and ϵ and ϵ' are the correct and indicated strains, respectively. Equation (3.12) clearly shows that error will be introduced in the strain readings. If the attenuation is to be less than 0.5%, R_L/R_g should be less than 0.0025. Thus for a 120-Ω gage, R_L should be less than 0.30 Ω.

To compensate for the error introduced by the lead wires, the gage factor should be set at G_f', which is calculated by the following equation:

$$G_f' = \frac{G_f}{1 + 2R_L/R_g}$$

where G_f is the gage factor given by the manufacturer.

EXAMPLE 3.8

If each lead in Fig. 3.8 is a 25-ft-long copper wire with a 0.03-in. diameter, what is the total resistance of the leads? If a strain gage with a resistance of 120-Ω and a gage factor of 2.16 is used, what modified gage factor should be used in a strain indicator in order to read directly the "correct" strains of the specimen? The copper has a resistance of 0.017-Ω per meter of length for a 1-mm^2 conductor.

Solution   1 m = 3.28 ft, 1 mm^2 = 0.00155 in^2. The cross-sectional area of the copper lead is given by

$$A = \frac{\pi D^2}{4} = \frac{\pi(0.03)^2}{4} = 0.000707 \text{ in}^2$$

Thus

$$R_L = \frac{(0.017)(25/3.28)}{0.000707/0.00155} = 0.284 \, \Omega$$

The total resistance of the leads is $0.568 \, \Omega$:

$$G'_f = \frac{G_f}{1 + 2R_L/R_g} = \frac{2.16}{1 + 0.568/120} = 2.15$$

Therefore, the gage factor should be set at 2.15 on the strain indicator.

Note that if a 350-Ω strain gage is used, the error introduced by the leads will be reduced. Using a larger-resistance strain gage will also be helpful in initially balancing the bridge.

Temperature compensation can be lost if long lead wires are used with the connection shown in Fig. 3.8. Suppose that gage 4 is the dummy gage used to compensate the temperature effect. Assume that gage 1, gage 4, and the lead wires are subjected to temperature change ΔT when the strain is being measured by gage 1. The out-of-balance voltage can be expressed as

$$\Delta E = V \frac{m}{(1 + m)^2} \left(\frac{\Delta R_{g\epsilon}}{R_1} + \frac{\Delta R_{g\Delta T}}{R_1} + \frac{2\Delta R_{L\Delta T}}{R_1} - \frac{\Delta R_{4\Delta T}}{R_4} \right) \tag{3.13}$$

where $R_1 = R_g + 2R_L$ is the resistance of arm 1 and $R_4 = R_g$ is the resistance of the dummy gage with negligible resistance of the lead wires. These four terms represent the resistance change in the active gage due to the applied load, the resistance change in the active gage due to temperature change ΔT, the resistance change in the long lead wires due to temperature change ΔT, and the resistance change in the dummy gage due to temperature change ΔT. It is clear that temperature compensation cannot be achieved since in general the sum of the second, the third, and the fourth terms cannot be zero.

EXAMPLE 3.9

Suppose that copper has a temperature coefficient of resistivity of $0.0022 \, °F^{-1}$; calculate the apparent strain component due to temperature change of $1°F$ in the lead wires for Example 3.8.

Solution   Due to the $1°F$ temperature change, the total resistance change of the lead wires is

$$\Delta R_L = 0.568 \times 0.0022 = 0.00125 \, \Omega$$

Thus

$$\epsilon_{app} = \frac{\Delta R_L}{RG_f} = \frac{0.00125}{(120)(2.16)} = 4.82\mu\epsilon \text{ per } °F$$

To achieve temperature compensation, a three-lead-wire connection, as shown in Fig. 3.9, should be employed when the active gage and compensating gage (either dummy gage or another active gage) can be placed adjacently. The three-lead-wire sys-

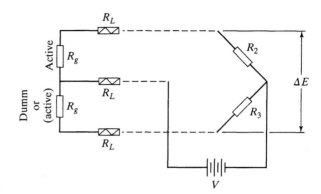

FIGURE 3.9   Three-lead-wire arrangement.

tem may still have a problem of signal attenuation; however, this attenuation is reduced by a factor of approximately 2 compared with the two-lead-wire system. If the adjusted gage factor G_f' is used to eliminate the error, it can be shown that G_f' is given by the equation

$$G_f' = \frac{G_f}{1 + R_L/R_g}$$

It should be pointed out that G_f' is the same regardless of whether gage 4 is a dummy gage or an active gage.

To show that temperature compensation can be achieved in a three-lead-wire system, let's consider the case when gage 4 is a dummy gage. Equation (3.13) then becomes

$$\Delta E = \frac{Vm}{(1 + m)^2}\left(\frac{\Delta R_{g\epsilon}}{R_1} + \frac{\Delta R_{g\Delta T}}{R_1} + \frac{2\Delta R_{L\Delta T}}{R_1} - \frac{\Delta R_{g\Delta T}}{R_4} - \frac{2\Delta R_{L\Delta T}}{R_4}\right) \quad (3.14)$$

It is obvious that the sum of the terms related to temperature change is zero if $R_1 = R_4 = R_g + 2R_L$ and if the gages and lead wires are subjected to exactly the same temperature change.

3.4 STRAIN GAGE TRANSDUCERS

A strain gage transducer is a device that uses strain gages as the sensor to produce an electrical signal that is directly proportional to such mechanical quantities as force, displacement, pressure, torque, and acceleration. In principle, Examples 3.5, 3.6, and 3.7 are applications of strain gages for transducers. Many different types of sophisticated strain gage transducers are commercially available. A few transducers—namely, load cell, torque meter, accelerometer, and displacement transducer—are shown in Fig. 3.10. The operating principles of a load cell and torque meter are described in this section.

Figure 3.10a shows a tension–compression load cell. To increase sensitivity and accuracy (to eliminate possible bending or torsional effects), four strain gages are mounted on the central region of the bar with two gages in the axial direction and two

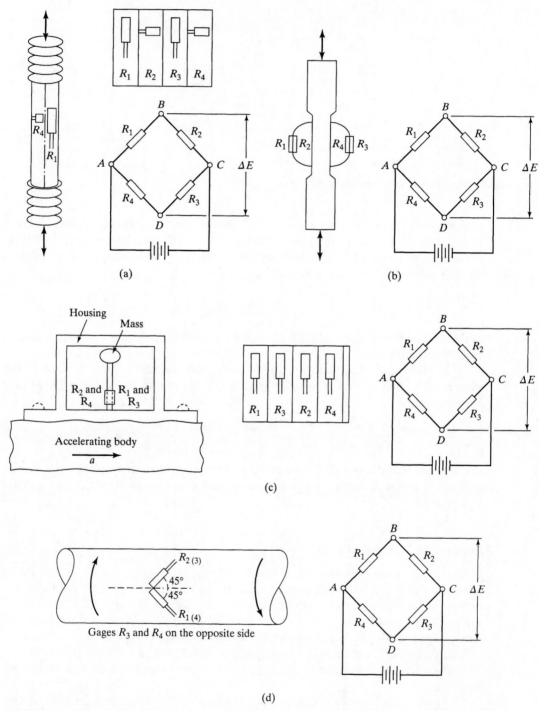

FIGURE 3.10   Selected examples of strain gage transducers: (a) tension–compression load cell; (b) extensometer; (c) accelerometer; (d) torque meter.

gages in the transverse direction and connected in a full bridge; the two gages in each set are bonded at diametrically opposite locations. When a load P is applied to the bar, the axial and transverse strains are

$$\epsilon_a = \frac{P}{EA} \qquad \epsilon_t = -\frac{\mu P}{EA} = -\mu\epsilon_a \tag{3.15}$$

where E is Young's modulus of the bar, A the cross-sectional area of the member, and μ is Poisson's ratio.

If arms R_1 and R_3 are gages mounted in the axial direction, and arms R_2 and R_4 are gages mounted in the transverse direction, the changes of resistance in four gages can be determined by using Eq. (3.15):

$$\frac{\Delta R_1}{R_1} = \frac{\Delta R_3}{R_3} = G_f\epsilon_a = \frac{G_f P}{EA} \qquad \frac{\Delta R_2}{R_2} = \frac{\Delta R_4}{R_4} = G_f\epsilon_t = -\frac{\mu G_f P}{EA} \tag{3.16}$$

Substituting Eq. (3.16) into Eq. (3.4) and noting that $m = 1$ gives

$$\Delta E = \frac{(1 + \mu)VG_f}{2EA}P$$

From the equation above, it is clear that the relation between out-of-balance voltage and applied load P is linear if the deformation of the bar is elastic, so it can be calibrated to read the load directly.

To convert a cylindrical piece into a torque meter, four gages A, B, C, and D are mounted on the center portion of the piece with two gages (A and C) oriented at an angle 45° with respect to the axis of the shaft (the x axis), and two gages (B and D) oriented at an angle −45° with respect to the axis of the shaft, as shown in Fig. 3.10d. Consider two strain gages A and B; the strains along the gage axes are obtained by using Eq. (1.34a) as

$$\epsilon_A = \frac{\epsilon_x + \epsilon_y}{2} + \frac{\gamma_{xy}}{2}$$

$$\epsilon_B = \frac{\epsilon_x + \epsilon_y}{2} - \frac{\gamma_{xy}}{2}$$

From the equations above, the shear strain γ_{xy} is

$$\gamma_{xy} = \epsilon_A - \epsilon_B = \frac{\tau_{xy}}{G} = \frac{Tr}{GJ} \tag{3.17}$$

where G is the shear modulus of the material of the structural member, T the applied torque, r the radius of the cylindrical member, and J the centroidal polar moment of inertia of the cross section, respectively.

Let arms R_1 and R_3 be gages A and C, and arms R_2 and R_4 be gages B and D. The changes of resistance in the four gages are

$$\frac{\Delta R_1}{R_1} = \frac{\Delta R_3}{R_3} = G_f\epsilon_A \qquad \frac{\Delta R_2}{R_2} = \frac{\Delta R_4}{R_4} = G_f\epsilon_B \tag{3.18}$$

Substituting Eq. (3.17) and Eq. (3.18) into Eq. (3.4) and noting that $m = 1$ gives

$$\Delta E = \frac{V G_f r}{2GJ} T$$

It is obvious that the relation between out-of-balance voltage and applied torque T is linear if the deformation of the shaft is elastic. Note that the device is essentially independent of the effects of axial and bending strains, so it is often called a *torque meter*.

3.5   STRAIN GAGE ROSETTE DATA ANALYSIS AND CORRECTION

Strain gages are normally used to determine the plane stress states at a point on the free surface of a structural or machine part subjected to applied loads. Let the x and y axes be in the plane of free surface and the z axis be perpendicular to it; then $\sigma_z = \tau_{zx} = \tau_{zy} = 0$. In general, a three-element rosette is required to determine the stresses σ_x, σ_y, and τ_{xy}, since the directions of the principal stresses are normally not known. Thus there are three independent unknowns: σ_x, σ_y, and τ_{xy}; alternatively, the unknowns are the principal stresses σ_1 and σ_2 and the principal angle α_1. In some cases the directions of the principal stresses are known beforehand, and then only a two-element rectangular rosette is needed. The rosette is bonded to the surface so that each gage is lined up with a principal direction.

Consider a general three-element rosette, whose orientations are shown in Fig. 3.11. By using the equations of strain transformation [i.e., Eq. (1.34)], we obtain

$$\epsilon_I = \epsilon_x \cos^2 \beta_I + \epsilon_y \sin^2 \beta_I + \gamma_{xy} \sin \beta_I \cos \beta_I$$

$$\epsilon_{II} = \epsilon_x \cos^2 \beta_{II} + \epsilon_y \sin^2 \beta_{II} + \gamma_{xy} \sin \beta_{II} \cos \beta_{II} \qquad (3.19)$$

$$\epsilon_{III} = \epsilon_x \cos^2 \beta_{III} + \epsilon_y \sin^2 \beta_{III} + \gamma_{xy} \sin \beta_{III} \cos \beta_{III}$$

For a given set of angles β_I, β_{II}, and β_{III}, the strains ϵ_x, ϵ_y, and γ_{xy} can be obtained by solving Eqs. (3.19) in terms of the measured strains ϵ_I, ϵ_{II}, and ϵ_{III}. Therefore, the principal strains ϵ_1 and ϵ_2 and the principal angle α_1 (between ϵ_1 and the x axis) can be obtained by Eqs. (1.37) and (1.38), which are rewritten as

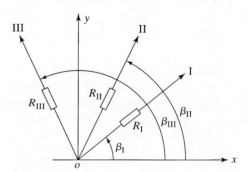

FIGURE 3.11   Orientation of a three-element rosette.

$$\epsilon_1 = \frac{1}{2}(\epsilon_x + \epsilon_y) + \frac{1}{2}[(\epsilon_x - \epsilon_y)^2 + \gamma_{xy}^2]^{1/2}$$

$$\epsilon_2 = \frac{1}{2}(\epsilon_x + \epsilon_y) - \frac{1}{2}[(\epsilon_x - \epsilon_y)^2 + \gamma_{xy}^2]^{1/2} \qquad (3.20)$$

$$\alpha_{1(2)} = \frac{1}{2}\tan^{-1}\frac{\gamma_{xy}}{\epsilon_x - \epsilon_y}$$

or

$$\alpha_1 = \tan^{-1}\frac{\gamma_{xy}}{2(\epsilon_1 - \epsilon_y)} = \tan^{-1}\frac{2(\epsilon_1 - \epsilon_x)}{\gamma_{xy}}$$

Once the principal strains ϵ_1 and ϵ_2, Young's modulus E, and Poisson's ratio μ are known, the principal stresses σ_1 and σ_2 can be found by replacing the subscripts x and y in Eq. (1.47) by 1 and 2: namely,

$$\sigma_1 = \frac{E}{1 - \mu^2}(\epsilon_1 + \mu\epsilon_2)$$

$$\sigma_2 = \frac{E}{1 - \mu^2}(\epsilon_2 + \mu\epsilon_1) \qquad (3.21)$$

Note that the directions of the principal stresses are the same as those of the principal strains.

EXAMPLE 3.10 [7]

By considering the equilibrium of the free-body diagram, shown in Fig. E3.10, derive an expression to compute the principal angle α_1 in terms of principal stress and Cartesian stress components at a point. Then prove that the principal angle can be conveniently computed by the following equation [the last equation in Eq. (3.20)]:

$$\alpha_i = \tan^{-1}\frac{2(\epsilon_i - \epsilon_x)}{\gamma_{xy}} = \tan^{-1}\frac{\gamma_{xy}}{2(\epsilon_i - \epsilon_y)}$$

where α_i, ϵ_i ($i = 1,2$), ϵ_x, ϵ_y, and γ_{xy} are the principal angle, principal strain, strain in the x direction, strain in the y direction, and engineering shear strain, respectively.

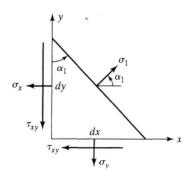

FIGURE E3.10

Solution Without loss of generality, we assume that the dimension of the free body in the z direction equals 1. The equations of equilibrium of the free body in the x and y directions are

$$\Sigma F_x = \sigma_i \frac{dy}{\cos \alpha_i} \cos \alpha_i - \sigma_x \, dy - \tau_{xy} \, dx = 0$$

$$\Sigma F_y = \sigma_i \frac{dx}{\sin \alpha_i} \sin \alpha_i - \sigma_y \, dx - \tau_{xy} \, dy = 0$$

Since $\tan \alpha_i = dx/dy$, thus

$$\tan \alpha_i = \frac{\sigma_i - \sigma_x}{\tau_{xy}} = \frac{\tau_{xy}}{\sigma_i - \sigma_y}$$

or

$$\alpha_i = \tan^{-1} \frac{\sigma_i - \sigma_x}{\tau_{xy}} = \tan^{-1} \frac{\tau_{xy}}{\sigma_i - \sigma_y}$$

By replacing σ_i, σ_x, σ_y, and τ_{xy} with ϵ_i, ϵ_x, ϵ_y, and $\gamma_{xy}/2$ in the above equation, we obtain

$$\tan \alpha_i = \frac{(\epsilon_i - \epsilon_x)}{\gamma_{xy}} = \frac{\gamma_{xy}}{2(\epsilon_i - \epsilon_y)}$$

or

$$\alpha_i = \tan^{-1} \frac{2(\epsilon_i - \epsilon_x)}{\gamma_{xy}} = \tan^{-1} \frac{\gamma_{xy}}{2(\epsilon_i - \epsilon_y)}$$

The two most commonly used three-element rosettes are the rectangular rosette and the delta rosette. Consider first the three-element rectangular rosette. For mere convenience, let the rosette be positioned at the angles of 0°, 45°, and 90°, as shown in Fig. 3.12.

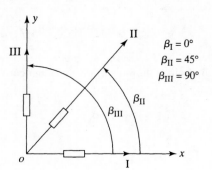

FIGURE 3.12 Gage orientations of a three-element rectangular rosette.

With $\beta_I = 0°$, $\beta_{II} = 45°$, and $\beta_{III} = 90°$, Eq. (3.19) becomes

$$\epsilon_I = \epsilon_x$$

$$\epsilon_{II} = \frac{\epsilon_x + \epsilon_y + \gamma_{xy}}{2}$$

$$\epsilon_{III} = \epsilon_y$$

(3.22)

Solving the equations above shows that the shear strain is

$$\gamma_{xy} = 2\epsilon_{II} - \epsilon_I - \epsilon_{III} \tag{3.23}$$

Substituting Eqs. (3.22) and (3.23) into Eq. (3.20), the principal strains and the principal angle can be written as

$$\epsilon_1 = \frac{1}{2}(\epsilon_I + \epsilon_{III}) + \frac{1}{2}[(\epsilon_I - \epsilon_{III})^2 + (2\epsilon_{II} - \epsilon_I - \epsilon_{III})^2]^{1/2}$$

$$\epsilon_2 = \frac{1}{2}(\epsilon_I + \epsilon_{III}) - \frac{1}{2}[(\epsilon_I - \epsilon_{III})^2 + (2\epsilon_{II} - \epsilon_I - \epsilon_{III})^2]^{1/2} \tag{3.24}$$

$$\alpha_1 = \tan^{-1}\frac{2\epsilon_{II} - \epsilon_I - \epsilon_{III}}{2(\epsilon_1 - \epsilon_{III})} = \tan^{-1}\frac{2(\epsilon_1 - \epsilon_I)}{2\epsilon_{II} - \epsilon_I - \epsilon_{III}}$$

$$\alpha_2 = \alpha_1 + 90°$$

Substituting Eq. (3.24) into Eq. (3.21), the corresponding principal stresses can be expressed in terms of the measured strains as follows:

$$\sigma_1 = E\left\{\frac{\epsilon_I + \epsilon_{III}}{2(1 - \mu)} + \frac{1}{2(1 + \mu)}[(\epsilon_I - \epsilon_{III})^2 + (2\epsilon_{II} - \epsilon_I - \epsilon_{III})^2]^{1/2}\right\}$$

$$\sigma_2 = E\left\{\frac{\epsilon_I + \epsilon_{III}}{2(1 - \mu)} - \frac{1}{2(1 + \mu)}[(\epsilon_I - \epsilon_{III})^2 + (2\epsilon_{II} - \epsilon_I - \epsilon_{III})^2]^{1/2}\right\} \tag{3.25}$$

EXAMPLE 3.11

A rectangular three-element rosette was mounted on an aluminum structure. The three strain readings are $\epsilon_I = 400\mu\epsilon$, $\epsilon_{II} = 1800\mu\epsilon$, and $\epsilon_{III} = 1200\mu\epsilon$. Find the values and directions of the principal strains and the values of the principal stresses. Young's modulus and Poisson's ratio are 10,000 ksi and 0.33, respectively.

Solution

$$\epsilon_I + \epsilon_{III} = 1600\mu\epsilon$$

$$\epsilon_I - \epsilon_{III} = -800\mu\epsilon$$

$$2\epsilon_{II} - \epsilon_I - \epsilon_{III} = (2 \times 1800) - 400 - 1200 = 2000\mu\epsilon$$

$$0.5[(\epsilon_2 - \epsilon_{III})^2 + (2\epsilon_{II} - \epsilon_I - \epsilon_{III})^2]^{1/2} = 0.5[(-800)^2 + (2000)^2]^{1/2} = 1077\mu\epsilon$$

Using Eq. (3.24), we get

$$\epsilon_1 = \frac{1600}{2} + 1077 = 1877\mu\epsilon$$

$$\epsilon_2 = \frac{1600}{2} - 1077 = -277\mu\epsilon$$

$$\alpha_1 = \tan^{-1}\frac{2000}{2(1877 - 1200)} = 56°$$

$$\alpha_2 = \alpha_1 + 90° = 146°$$

The principal stresses σ_1 and σ_2 are computed by Eq. (3.24); thus

$$\sigma_1 = 10,000 \left[\frac{1600}{2(1 - 0.33)} + \frac{1077}{(1 + 0.33)} \right] (10^{-6}) = 20.0 \text{ ksi}$$

$$\sigma_2 = 10,000 \left[\frac{1600}{2(1 - 0.33)} - \frac{1077}{(1 + 0.33)} \right] (10^{-6}) = 3.84 \text{ ksi}$$

Consider next the delta rosette, which is oriented at angles of 0°, 60°, and 120°, as shown in Fig. 3.13. With $\beta_I = 0°$, $\beta_{II} = 120°$, and $\beta_{III} = 240°$, Eq. (3.19) becomes

$$\epsilon_x = \epsilon_I$$

$$\epsilon_y = \frac{1}{3} [2(\epsilon_{II} + \epsilon_{III}) - \epsilon_I]$$ (3.26)

$$\gamma_{xy} = \frac{2\sqrt{3}}{3} (\epsilon_{III} - \epsilon_{II})$$

Therefore, the principal strains and the principal angles can be expressed as

$$\epsilon_1 = \frac{\epsilon_I + \epsilon_{II} + \epsilon_{III}}{3} + \left\{ \left[\epsilon_I - \frac{\epsilon_I + \epsilon_{II} + \epsilon_{III}}{3} \right]^2 + \frac{(\epsilon_{II} - \epsilon_{III})^2}{3} \right\}^{1/2}$$

$$\epsilon_2 = \frac{\epsilon_I + \epsilon_{II} + \epsilon_{III}}{3} - \left\{ \left[\epsilon_I - \frac{\epsilon_I + \epsilon_{II} + \epsilon_{III}}{3} \right]^2 + \frac{(\epsilon_{II} - \epsilon_{III})^2}{3} \right\}^{1/2}$$ (3.27)

$$\alpha_1 = \tan^{-1} \frac{\sqrt{3}(\epsilon_1 - \epsilon_I)}{\epsilon_{III} - \epsilon_{II}} \qquad \alpha_2 = \alpha_1 + 90°$$

The corresponding principal stresses are

$$\sigma_1 = E \left[\frac{\epsilon_I + \epsilon_{II} + \epsilon_{III}}{3(1 - \mu)} + \frac{1}{1 + \mu} \left\{ \left[\epsilon_I - \frac{\epsilon_I + \epsilon_{II} + \epsilon_{III}}{3} \right]^2 + \frac{(\epsilon_{II} - \epsilon_{III})^2}{3} \right\}^{1/2} \right]$$

$$\sigma_2 = E \left[\frac{\epsilon_I + \epsilon_{II} + \epsilon_{III}}{3(1 - \mu)} - \frac{1}{1 + \mu} \left\{ \left[\epsilon_I - \frac{\epsilon_I + \epsilon_{II} + \epsilon_{III}}{3} \right]^2 + \frac{(\epsilon_{II} - \epsilon_{III})^2}{3} \right\}^{1/2} \right]$$ (3.28)

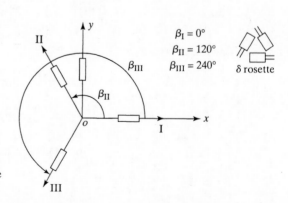

$$\beta_I = 0°$$
$$\beta_{II} = 120°$$
$$\beta_{III} = 240°$$

δ rosette

FIGURE 3.13   Gage orientations of the delta rosette.

For convenience, a summary of the analytical expressions for four commonly employed rosettes is presented in Table 3.1.

In the discussion above, it is assumed that any necessary corrections, such as those for gage factor and for transverse sensitivity of the strain gages, have been made. In Section 2.5, the general equation, Eq. (2.18), was derived for correcting transverse effects in biaxial strain state. It can also be employed directly for the two-element rectangular rosette. For other rosettes, however, new equations are necessary. In Example 3.12, the equations for correcting transverse sensitivity in a three-element rectangular rosette will be derived.

EXAMPLE 3.12

Show that the equations for correcting transverse sensitivity in a three-element rectangular rosette, as shown in Fig. E3.12a, are given by

$$\epsilon_I = Q(\epsilon_I' - K_t\epsilon_{III}')$$
$$\epsilon_{II} = Q[(1 + K_t)\epsilon_{II}' - K_t(\epsilon_I' + \epsilon_{III}')]$$
$$\epsilon_{III} = Q(\epsilon_{III}' - K_t\epsilon_I')$$

where ϵ_I', ϵ_{II}', and ϵ_{III}' are indicated strains and ϵ_I, ϵ_{II}, and ϵ_{III} are corrected strains; K_t is the transverse sensitivity and is assumed the same for all three elements, $Q = (1 - \mu_c K_t)/(1 - k_t^2)$; and μ_c is Poisson's ratio, which is used by the manufacturers for gage factor and transverse sensitivity factor determinations (usually, $\mu_c = 0.285$).

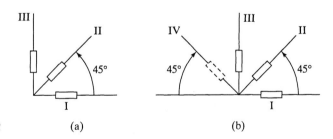

FIGURE E3.12   (a)   (b)

Solution   Using Eq. (2.20), we obtain directly

$$\epsilon_I = Q(\epsilon_I' - K_t\epsilon_{III}')$$
$$\epsilon_{III} = Q(\epsilon_{III}' - K_t\epsilon_I')$$

Imagine that there were another gage IV, mounted in the direction perpendicular to gage II, as shown in Fig. E3.12b. Thus, using Eq. (2.20),

$$\epsilon_{II} = Q(\epsilon_{II}' - K_t\epsilon_{IV}')$$
$$\epsilon_{IV} = Q(\epsilon_{IV}' - K_t\epsilon_{II}')$$

According to the theory of elasticity, the sum of strains in any two perpendicular directions is an invariant, so we have

$$\epsilon_I + \epsilon_{III} = \epsilon_{II} + \epsilon_{IV}$$

TABLE 3.1 Summary of Expressions Among Strain Rosette Readings and Principal Strains, Stresses, and Angles[a]

Rosette Type	Gage Pattern	Principal Strains and Stresses	Principal Angle α_1
Two-element rectangular rosette		$\epsilon_1 = \epsilon_I, \quad \epsilon_2 = \epsilon_{II}$ $\sigma_1 = \dfrac{E}{1-\mu^2}(\epsilon_I + \mu\epsilon_{II})$ $\sigma_2 = \dfrac{E}{1-\mu^2}(\epsilon_{II} + \mu\epsilon_I)$	$\alpha_1 = 0, \quad \epsilon_I > \epsilon_{II},$ $\left(\alpha_1 = \dfrac{\pi}{2}, \quad \epsilon_I < \epsilon_{II}\right)$ $\alpha_2 = \alpha_1 + \dfrac{\pi}{2}$
Three-element rectangular rosette		$\epsilon_{1,2} = \dfrac{\epsilon_I + \epsilon_{III}}{2} \pm \dfrac{1}{2}\sqrt{(\epsilon_I - \epsilon_{III})^2 + (2\epsilon_{II} - \epsilon_I - \epsilon_{III})^2}$ $\sigma_{1,2} = \dfrac{E}{2}\left[\dfrac{\epsilon_I + \epsilon_{III}}{1-\mu} \pm \dfrac{1}{1+\mu}\sqrt{(\epsilon_I - \epsilon_{III})^2 + (2\epsilon_{II} - \epsilon_I - \epsilon_{III})^2}\right]$	$\alpha_1 = \tan^{-1}\dfrac{2\epsilon_{II} - \epsilon_I - \epsilon_{III}}{2(\epsilon_I - \epsilon_{III})}$ $\alpha_2 = \alpha_1 + \dfrac{\pi}{2}$
Delta rosette		$\epsilon_{1,2} = \dfrac{\epsilon_I + \epsilon_{II} + \epsilon_{III}}{3} \pm \dfrac{1}{3}\sqrt{\{2\epsilon_I - (\epsilon_{II} + \epsilon_{III})\}^2 + 3(\epsilon_{II} - \epsilon_{III})^2}$ $\sigma_{1,2} = \dfrac{E}{3}\left[\dfrac{\epsilon_I + \epsilon_{II} + \epsilon_{III}}{1-\mu} \pm \dfrac{1}{3(1+\mu)}\sqrt{\{2\epsilon_I - (\epsilon_{II} + \epsilon_{III})\}^2 + 3(\epsilon_{II} - \epsilon_{III})^2}\right]$	$\alpha_1 = \tan^{-1}\dfrac{\sqrt{3}(\epsilon_I - \epsilon_I)}{\epsilon_{III} - \epsilon_{II}}$ $\alpha_2 = \alpha_1 + \dfrac{\pi}{2}$

[a]E, Young's modulus; μ, Poisson's ratio.

TABLE 3.2 Summary of Expressions for Correcting Transverse Strain Effects for various Rosettes[a]

Type of Rosette		
Name	Gage Pattern	Expression for Correction
Two-element rectangular rosette		$\epsilon_I = Q(\epsilon'_I - K_t \epsilon'_{II})$ $\epsilon_{II} = Q(\epsilon'_{II} - K_t \epsilon'_I)$
Three-element rectangular rosette		$\epsilon_I = Q(\epsilon'_I - K_t \epsilon'_{III})$ $\epsilon_{II} = Q[(1 + K_t)\,\epsilon'_{II} - K_t\,(\epsilon'_I + \epsilon'_{III})]$ $\epsilon_{III} = Q(\epsilon'_{III} - K_t \epsilon'_I)$
Delta rosette		$\epsilon_I = Q\left[\left(1 + \dfrac{K_t}{3}\right)\epsilon'_I - \dfrac{2K_t}{3}(\epsilon'_{II} + \epsilon'_{III})\right]$ $\epsilon_{II} = Q\left[\left(1 + \dfrac{K_t}{3}\right)\epsilon'_{II} - \dfrac{2K_t}{3}(\epsilon'_{III} + \epsilon'_I)\right]$ $\epsilon_{III} = Q\left[\left(1 + \dfrac{K_t}{3}\right)\epsilon'_{III} - \dfrac{2K_t}{3}(\epsilon'_I + \epsilon'_{II})\right]$

[a]ϵ'_i, indicated strain; ϵ_i, corrected strain; $Q = (1 - \mu_c K_t)/(1 - K_t^2)$.

and

$$\epsilon'_I + \epsilon'_{III} = \epsilon'_{II} + \epsilon'_{IV}$$

Therefore,

$$\epsilon'_{IV} = \epsilon'_I + \epsilon'_{III} - \epsilon'_{II}$$
$$\epsilon_{II} = Q(\epsilon'_{II} - K_t \epsilon'_{IV}) = Q[\epsilon'_{II} - K_t(\epsilon'_I + \epsilon'_{III} - \epsilon'_{II})]$$

or

$$\epsilon_{II} = Q[(1 + K_t)\,\epsilon'_{II} - K_t(\epsilon'_I + \epsilon'_{III})]$$

By the same procedures, all equations in Table 3.2 can easily be obtained. The key step is to obtain the strain invariant $(\epsilon'_1 + \epsilon'_2)$ for various rosettes.

EXAMPLE 3.13

Assume that the transverse sensitivity factor $K_t = 2.1\%$ for the rectangular rosette in Example 3.11. Calculate the principal strains and the principal stresses after correction of the transverse sensitivity effects.

Solution By using expressions for the three-element rectangular rosette in Table 3.2, one obtains

$$Q = \frac{1 - (0.285)(0.021)}{1 - (0.021)^2} = 0.994$$

$$\epsilon_I = 0.994[400 - (0.021)(1200)] = 373\mu\epsilon$$

$$\epsilon_{II} = 0.994[(1 + 0.021)(1800) - (0.021)(400 + 1200)] = 1794\mu\epsilon$$

$$\epsilon_{III} = 0.994[1200 - (0.021)(400)] = 1185\mu\epsilon$$

$$\epsilon_I + \epsilon_{III} = 1558\mu\epsilon$$

$$\epsilon_I - \epsilon_{III} = -812\mu\epsilon$$

$$2\epsilon_{II} - \epsilon_I - \epsilon_{III} = 2(1794) - 1558 = 2030\mu\epsilon$$

$$\frac{[(\epsilon_I - \epsilon_{III})^2 + (2\epsilon_{II} - \epsilon_I - \epsilon_{III})^2]^{1/2}}{2} = \frac{[(-812)^2 + (2030)^2]^{1/2}}{2} = 1093\mu\epsilon$$

Equation (3.24) gives

$$\epsilon_1 = \frac{1558}{2} + 1093 = 1872\mu\epsilon$$

$$\epsilon_2 = \frac{1558}{2} - 1093 = -314\mu\epsilon$$

$$\alpha_1 = \tan^{-1}\frac{2(1872 - 373)}{2030} = 56°$$

$$\alpha_2 = \alpha_1 + 90° = 146°$$

Eq. (3.25) gives

$$\sigma_1 = 10,000\left[\frac{1558}{2(1 - 0.33)} + \frac{1093}{1 + 0.33}\right](10^{-6}) = 19.84 \text{ ksi}$$

$$\sigma_2 = 10,000\left[\frac{1558}{2(1 - 0.33)} - \frac{1093}{1 + 0.33}\right](10^{-6}) = 3.41 \text{ ksi}$$

Similarly, equations for correcting transverse sensitivity for the delta rosette can also be obtained. For completeness, a summary of expressions to correct transverse effects for three commonly used rosettes is presented in Table 3.2. It is pointed out that the transverse sensitivity factor is assumed to be the same for all elements in Table 3.2. If it is not true, the reader is referred to, for example, reference [5] for details.

3.6 CORRECTION FOR WHEATSTONE BRIDGE NONLINEARITY

As mentioned earlier, a nonlinearity effect must be considered when large strains (e.g., 5% or greater) are to be measured by using an unbalanced Wheatstone bridge. For a large class of commercial static strain indicators and signal conditioners employing an unbalanced Wheatstone bridge, it is necessary to provide a simple means for correcting the error due to the Wheatstone bridge nonlinearity [3].

Consider a general Wheatstone bridge, as shown in Fig. 3.14. Assume that gage R_1 is subjected to a strain ϵ, gage R_2 to a strain $a\epsilon$, gage R_3 to a strain $b\epsilon$, and gage R_4 to a strain $c\epsilon$. In practice, $a = 0, -\mu$, or -1; $b = 0, \mu$, or 1; and $c = 0, -\mu$, or -1. Without loss of generality, we assume that the bridge is initially balanced and $R_1 = R_2 = R_3 = R_4 = R$. Furthermore, we assume that G_f is the gage factor set on the strain indicator (or signal conditioner); thus $\Delta R_1/R_1 = G_f\epsilon$, $\Delta R_2/R_2 = aG_f\epsilon$, $\Delta R_3/R_3 = bG_f\epsilon$, and $\Delta R_4/R_4 = cG_f\epsilon$. The bridge nonlinearity is usually neglected in commercial strain indicators and signal conditioners and Eq. (3.4) is used to obtain the indicated strain ϵ^*, which differs from the actual strain ϵ when large strains are involved. Since $m = 1$, Eq. (3.4) becomes

$$\Delta E = \frac{V}{4}(1 - a + b - c)\, G_f\epsilon^* \tag{3.29}$$

or

$$(1 - a + b - c)\, \epsilon^* = \frac{4\Delta E}{VG_f} \tag{3.30}$$

where $(1 - a + b - c)\epsilon^*$ is the indicated strain read directly from the strain indicator.

However, the actual bridge output is given by

$$
\begin{aligned}
\frac{\Delta E}{V} &= \frac{(R_1 + \Delta R_1)(R_3 + \Delta R_3) - (R_2 + \Delta R_2)(R_4 + \Delta R_4)}{(R_1 + \Delta R_1 + R_2 + \Delta R_2)(R_3 + \Delta R_3 + R_4 + \Delta R_4)} \\[2mm]
&= \frac{(1 + \Delta R_1/R\,[1 + b\,(\Delta R_1/R)] - [1 + a\,(\Delta R_1/R)]\,[1 + c\,(\Delta R_1/R)]}{[2 + (1 + a)\,(\Delta R_1/R)]\,[2 + (b + c)\,(\Delta R_1/R)]} \\[2mm]
&= \frac{(1 - a + b - c)}{4}\frac{\Delta R_1}{R}\frac{4[(1 - a + b - c) + (b - ac)\,(\Delta R_1/R)]}{(1 - a + b - c)[4 + 2(1 + a + b + c)\,G_f\epsilon + (1 + a)(b + c)\,G_f^2\,\epsilon^2]}
\end{aligned}
$$

Substituting Eq. (3.31) into Eq. (3.30) gives

$$\epsilon^* = \frac{4[(1 - a + b - c) + (b - ac)G_f\epsilon]\epsilon}{(1 - a + b - c)\,[4 + 2(1 + a + b + c)G_f\epsilon + (1 + a)(b + c)(G_f\epsilon)^2]} \tag{3.32}$$

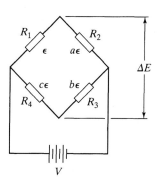

FIGURE 3.14   Typical Wheatstone bridge.

Equation (3.32) is a useful expression when corrections are necessary for Wheatstone bridge nonlinearity. Since a, b, and c are known beforehand in practice, the corrected strain ϵ can be obtained by solving Eq. (3.32). A few cases used frequently in practice are presented next.

Case 1: Quarter Bridge ($a = b = c = 0$). Only one active gage is used, such as in uniaxial tension or compression experiments. This is also the case frequently used in general two-dimensional strain measurements using strain gages. In this case, Eq. (3.32) is simplified as

$$\epsilon^* = \epsilon \frac{4}{4 + 2G_f\epsilon}$$

Solving the equation above for the corrected ϵ gives

$$\epsilon = \frac{2}{2 - G_f\epsilon^*} \epsilon^* \qquad (3.33)$$

Case 2: Half Bridge ($a = -\mu$, $b = c = 0$). Two active gages are used. This is the case in a uniaxial stress field with one gage in the uniaxial stress direction and the other in the transverse direction. The corrected strain ϵ can be computed using the equation,

$$\epsilon = \frac{2}{2 - (1 - \mu)G_f\epsilon^*} \epsilon^* \qquad (3.34)$$

Case 3: Full Bridge ($a = c = -\mu$, $b = 1$). Four active gages in uniaxial stress field are used, two gages are bonded along the uniaxial stress direction and two gages in transverse direction. The corrected strain ϵ is computed using the equation

$$\epsilon = \frac{2}{2 - (1 - \mu)G_f\epsilon^*} \epsilon^* \qquad (3.35)$$

There are three special cases ($a = -1, b = c = 0; a = -\mu, b = \mu, c = -1; a = c = -1, b = 1$) where the Wheatstone bridge output is always linear. In other words, no corrections are necessary no matter how large the measured strains are.

EXAMPLE 3.14

To obtain a uniaxial stress–strain curve for a material up to large strain at room temperature, a high-elongation strain gage was bonded to the specimen and a quarter bridge was used. At a certain tensile load, the indicated strain was 140,000$\mu\epsilon$. Determine the strain after correcting for the bridge nonlinearity and the error introduced if a correction had not been made. The gage factor set on the strain indicator was 2.17. If the load is compressive (the indicated strain is −140,000$\mu\epsilon$), what are the corrected strain ϵ and the error introduced if the correction is not made?

Solution Substituting $G_f = 2.17$ and $\epsilon^* = 0.14$ into Eq. (3.33) gives

$$\epsilon = \frac{(2)0.14}{2 - (2.17)(0.14)} = 0.165$$

$$\text{error} = \frac{0.165 - 0.14}{0.165}(100\%) = 15.2\%$$

When compressive load was applied, the indicated strain was negative. Substituting $G_f = 2.17$ and $\epsilon^* = -0.14$ into Eq. (3.33) gives

$$\epsilon = \frac{(2)(-0.14)}{2 - (2.17)(-0.14)} = -0.122$$

$$\text{error} = \frac{-0.122 - (-0.14)}{-0.122}(100\%) = -14.8\%$$

From the example above it is clear that the correction for the Wheatstone bridge nonlinearity is very important when large strains are to be measured by strain gages.

3.7 GAGE FACTOR FOR FINITE DEFORMATION

As mentioned earlier, gage factor variation is another major error source in high-elongation strain measurements by strain gages. It is a common practice to use $2 + \epsilon$ as the gage factor—correct in theory based on some simple assumptions but not, to our knowledge, substantiated by actual test results [4]. However, we found that the gage factor varies approximately according to the equation below [9]

$$G_f = C_1 + C_2\epsilon \tag{3.36}$$

where C_1 and C_2 are constants and ϵ is the nominal strain, respectively. C_1 is usually close to 2.0, but C_2 is usually greater than 1.0 and also close to 2.0. For example, $C_1 = 1.94$ and $C_2 = 1.82$ for a KFE-2-C1 gage, obtained by a uniaxial compression test on an annealed aluminum 1100 cylinder specimen and valid for strains larger than 0.04, as shown in Fig. 3.15.

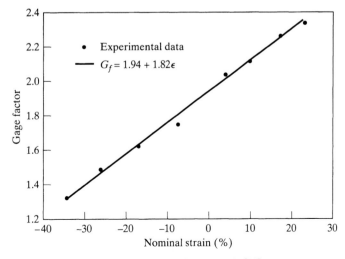

FIGURE 3.15   Gage factor variation with the strain [10].

It is difficult to provide a unique value for both constants C_1 and C_2, valid for any strain gages up to large strains. The values determined by experiments may depend on the gage materials, the lot number of the gages, the adhesive used to bond the strain gages, test conditions (tensile, compressive, or bending experiments), specimen material used, and specimen surface conditions (curved or flat) during the calibration.

3.8   SHUNT RESISTANCE CALIBRATION

Calibration of strain gage instrumentation—either instrument scaling or instrument verification [7]—is frequently done by indirect method. A resistance change (usually decreasing the resistance of a bridge arm) is produced by shunting a large resistor across one bridge arm to simulate a strain gage output. The basic shunt calibration of single active arm is shown in Fig. 3.16.

For simplicity and without loss generality, we assume that $R_1 = R_2 = R_3 = R_4$ and $R_1 = R_g$ (quarter bridge). Thus the bridge is initially balanced. By shunting R_2, a resistance decrease in bridge arm 2 is produced, which is given by

$$\Delta R_2 = \frac{R_2 R_c}{R_2 + R_c} - R_2 \quad \text{or} \quad \frac{\Delta R_2}{R_2} = \frac{-R_2}{R_2 + R_c} \tag{3.37}$$

where R_c is the calibration resistor. Since $R_2 = R_g$ and $\Delta R/R_g = G_f \epsilon$, we have

$$\epsilon_c = \frac{-R_g}{G_f(R_g + R_c)} \tag{3.38}$$

where ϵ_c is the strain simulated by shunting R_g with R_c. Equation (3.38) can be used to determine the value of calibration resistor for a given ϵ_c. For example, if $\epsilon_c = -5000\mu\epsilon$ and $G_f = 2$, $R_c = 11,880 \ \Omega$.

By using Eq. (3.4), it can be seen that a resistance decrease in bridge arm 2 by shunting a calibration resistor across R_2 will produce a positive output; in other words, a tensile strain is registered. For example, this calibration method is used in a P-3500 digital strain indicator [8].

It has been discussed in Section 3.3 that when the gage is remote from the instrument, the lead wire resistance must be taken into account. The usual way to compen-

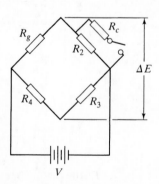

FIGURE 3.16   Basic shunt calibration Wheatstone bridge circuit.

sate for the error introduced by the lead wires is to adjust the gage factor. This can easily be done by shunt calibration. For example, if the measurement is to be done using a P-3500 digital strain indicator and a quarter bridge is to be used, corrections for any signal attenuation due to lead wire resistances can be done in the following steps:

1. Calculate ϵ' as follows:

$$\epsilon' = \frac{2}{G_f} \times 5000\mu\epsilon$$

where G_f is the gage factor provided by the manufacturer and $5000\mu\epsilon$ is the strain simulated by shunting R_c across R_2 $(= R_g)$ when the gage factor is set at 2.00.

2. Depress the CAL button.
3. Adjust the gage factor control knob until the strain indicator registers a strain ϵ'; then lock the knob.

Shunt calibration is simple in principle, but actually is more complicated. The reader is referred to reference [7] for more details.

3.9 POTENTIOMETER CIRCUIT

In dynamic strain gage applications, the potentiometer circuit is often useful, especially when the circuit output consists of two parts: one varies with time and the other is a constant. A typical potentiometer circuit is shown in Fig. 3.17. The voltage drops across R_1 or the voltage output of the potentiometer circuit, V_{AB}, is given by the equation

$$V_{AB} = \frac{R_1}{R_1 + R_2} V = \frac{V}{1 + m} \qquad (3.39)$$

where $m = R_2/R_1$ and V is the input voltage.

Consider the case that resistances R_1 and R_2 are changed by the amounts ΔR_1 and ΔR_2. The voltage output of the potentiometer circuit can be computed using Eq. (3.39), namely,

$$V_{AB} + \Delta V_{AB} = \frac{R_1 + \Delta R_1}{R_1 + \Delta R_1 + R_2 + \Delta R_2} V$$

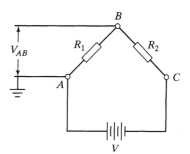

FIGURE 3.17 Typical potentiometer circuit.

If the nonlinearity of the potentiometer circuit is neglected (e.g., in the case when the strain to be measured is less than 1%), the output increment, ΔV_{AB}, can be expressed by the equation

$$\Delta V_{AB} = \frac{mV}{(1+m)^2} \left(\frac{\Delta R_1}{R_1} - \frac{\Delta R_2}{R_2} \right) \tag{3.40}$$

This is the basic equation governing the strain measurement using a potentiometer circuit.

The sensitivity of a potentiometer circuit, S_p, is defined by

$$S_p = \frac{\Delta V_{AB}}{\epsilon} = \frac{m}{(1+m)^2} \left(\frac{\Delta R_1}{R_1} - \frac{\Delta R_2}{R_2} \right) \frac{V}{\epsilon} \tag{3.41}$$

Since Eqs. (3.40) and (3.41) are exactly the same as those of an initially balanced half Wheatstone bridge, some previous discussions regarding quarter and half bridges, such as temperature compensation, sensitivity, lead wire effect, and nonlinearity, are equally valid for the potentiometer circuit. However, the actual output of a potentiometer circuit is $V_{AB} + \Delta V_{AB}$.

In general, ΔV_{AB} is much smaller than V_{AB} in magnitude. Thus it cannot be measured accurately when both ΔV_{AB} and V_{AB} are constant with respect to time; this is the case in the static strain measurement. However, in dynamic strain applications, ΔV_{AB} is changing with time but V_{AB} is a constant with respect to time. The constant component of the potentiometer circuit output, V_{AB}, can be blocked by using filters. In this way, ΔV_{AB} can be measured accurately.

PROBLEMS

3.1. Verify Eq. (3.3) if the second-order terms are neglected.

3.2. To increase the circuit sensitivity, the voltage can be varied in the quarter bridge, shown in Fig. P3.2. The active gage is a 350-Ω gage dissipating 0.05-W power with a gage factor of 2.12. Calculate the required voltage to be applied to the bridge and the resulting circuit sensitivity.

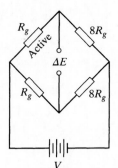

FIGURE P3.2

3.3. Prove that the bridge arrangement illustrated in Fig. 3.5 is temperature compensated.

3.4. The circuit sensitivity for the bridge arrangement, shown in Fig. 3.5, is given by Eq. (3.10). What could the theoretical value of maximum sensitivity be if the voltage were variable? ($\mu = 0.33$.)

3.5. Prove that the bridge arrangement shown in Fig. 3.7 is temperature compensated if all gages experience the same temperature change.

3.6. In addition to a bending moment M, a tensile load P is also applied on the beam shown in Fig. 3.7a. Show that the output ΔE is independent of the load P if a full bridge (Fig. 3.7b) is employed. Assume that the cross-sectional area of the beam is A.

3.7. To eliminate possible bending effect due to eccentric loading, to eliminate effect due to temperature change, and to increase the circuit sensitivity, four gages are mounted on a tensile specimen, as shown in Fig. P3.7. Determine the positions of the four gages in the bridge and derive an expression to relate the bridge output ΔE with the axial strain ϵ_a **(a)** if a half bridge is to be used, and **(b)** if a full bridge is employed. Assume that a fixed voltage V is applied to the bridge. The gage resistance, Poisson's ratio, and gage factor are R_g, μ, and G_f, respectively.

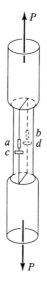

FIGURE P3.7

3.8. In Example 3.4, how should the four gages be placed in the bridge if a full Wheatstone bridge is to be used? What is the output ΔE in terms of V, G_f, μ, and ϵ?

3.9. Two quarter-circular gages (refer to Problem 2.7) are located in adjacent quadrants, as shown in Fig. P3.9. A half Wheatstone bridge is employed. Show that the bridge output $\Delta E/V$ is proportional to γ_{xy}/π. (Transverse sensitivity effects are neglected.)

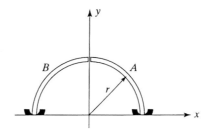

FIGURE P3.9

3.10. Four quarter-circular gages are located in all four quadrants, as shown in Fig. P3.10. A full Wheatstone bridge is used.

    (a) Position the four gages in the full bridge properly in order to have the bridge output proportional to shear strain γ_{xy}.

    (b) Determine the bridge output $\Delta E/V$. (Transverse sensitivity effects are neglected.)

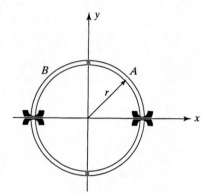

FIGURE P3.10

3.11. Two two-element shear gages (Fig. 2.3m or n) are mounted on a cylindrical piece at diametrically opposite locations, as shown in Fig. P3.11.

    (a) Position the four elements in a half Wheatstone bridge in order to have the bridge output proportional to shear strain (or torque T) and independent of bending moment M and axial load P.

    (b) Determine the bridge output $\Delta E/V$ and prove that it is independent of M and P. The cylindrical piece has a circular cross section with radius of R and a shear modulus of G.

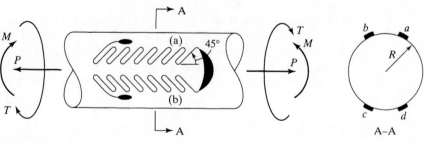

FIGURE P3.11

3.12. A cantilever beam with four strain gages is used as a load transducer, as shown in Fig. P3.12. A full Wheatstone bridge is to be used. Show the positions of four gages in the bridge so that maximum sensitivity is obtained and the output is equal to $(VG_f dh/2EI)P$; comment on the dimension d in relation to sensitivity. Note that V and G_f are the bridge excitation voltage and gage factor, respectively.

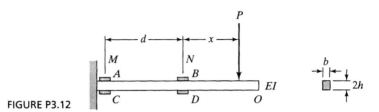

FIGURE P3.12

3.13. An extensometer, shown in Fig. 3.10b, consists of a semicircular-arch clip gage. The dimension of the clip gage is shown in Fig. P3.13, where $\Delta L = \dfrac{\pi F L^3}{16\, EI}$ and F is the force applied on the clip gage. Place gages 1 and 2 in a Wheatstone bridge and determine $\Delta E/V$ in terms of ϵ, G_f, t, and R when $\epsilon = \dfrac{\Delta L}{L}$ is the axial strain of the specimen and L is approximately the length of the clip gage.

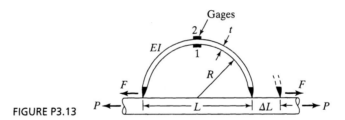

FIGURE P3.13

3.14. Determine ΔE of an accelerometer shown in Fig. 3.10C. The dimensions of the beam are shown in Fig. P3.14. Young's modulus of the beam is E.

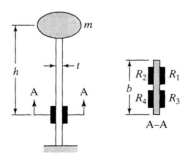

FIGURE P3.14

3.15. Using Mohr's strain circle, prove that the principal angle α_1 for the rectangular rosette, shown in Fig. 3.12, is

$$0° < \alpha_1 < 90° \qquad \text{when } \epsilon_{II} > \frac{\epsilon_I + \epsilon_{III}}{2}$$

$$-90° < \alpha_1 < 0° \qquad \text{when } \epsilon_{II} < \frac{\epsilon_I + \epsilon_{III}}{2}$$

$$\alpha_1 = 0° \qquad \text{when } \epsilon_I > \epsilon_{III} \quad \text{and} \quad \epsilon_{II} = \frac{\epsilon_I + \epsilon_{III}}{2}$$

$$\alpha_1 = \pm 90° \qquad \text{when } \epsilon_I < \epsilon_{III} \quad \text{and} \quad \epsilon_{II} = \frac{\epsilon_I + \epsilon_{III}}{2}$$

3.16. A three-element rectangular rosette is mounted on a steel pressure vessel at the region joining the spherical head and the cylindrical shell (Fig. P3.16). At a pressure of 500 psi, the strains indicated for elements 4, 5, and 6 are, respectively, $129\mu\epsilon$, $100\mu\epsilon$, and $123\mu\epsilon$. The gage factor is 2.08, but the value set on the strain indicator is 2.12. Determine the principal stresses and strains and the principal angle. Young's modulus for steel is 30,000 ksi. Poisson's ratio is 0.290.

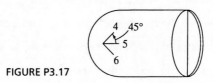

FIGURE P3.17

3.17. A delta rosette is mounted on a spherical head of a steel pressure vessel as shown in Fig. P3.17. At a pressure of 500 psi, the strains indicated for the three gages 1, 2, and 3 are, respectively, $64\mu\epsilon$, $74\mu\epsilon$, and $71\mu\epsilon$. Determine the principal stresses and strains and the principal angle. The gage factor is 2.12, which is set on the strain indicator. Young's modulus and Poisson's ratio for steel are 30,000 ksi and 0.290.

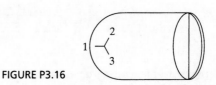

FIGURE P3.16

3.18. Determine expressions for principal strains and stresses for a T-delta rosette shown in Fig. P 3.18. Note that the advantage of adding a fourth gage is to check on the accuracy of the other three gages.

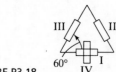

FIGURE P3.18

3.19. Prove that the equation for correcting the transverse sensitivity for element II in a T-delta rosette, shown in Fig. P3.18, is given by

$$\epsilon_{II} = Q[\epsilon'_{II} - K_t(\epsilon'_I + \epsilon'_{IV} - \epsilon'_{II})]$$

where $Q = (1 - \mu_c K_t)/(1 - K_t^2)$, ϵ'_i ($i =$ I, II, and IV) are indicated strain and ϵ_{II} is corrected strain, respectively. Assume that all elements have the same transverse sensitivity factor K_t.

3.20. Show that the expression for correcting transverse strain effects for element III in a delta rosette (listed in Table 3.2) is given by

$$\epsilon_{III} = Q\left[\left(1 + \frac{K_t}{3}\right)\epsilon'_{III} - \frac{2}{3} K_t (\epsilon'_I + \epsilon'_{II}) \right]$$

Assume that elements I, II, and III have the same transverse sensitivity factor K_t. $Q = (1 - \mu_c K_t)/(1 - K_t^2)$, ϵ'_i ($i =$ I, II, and III) are indicated strains, and ϵ_{III} is the corrected strain for element III.

3.21. In Problem 3.16, the factor of transverse sensitivity for the three-element rectangular rosette is +3.1%. Determine the principal stresses and the principal angle. Also determine the error in determining the maximum principal stresses if the effect of transverse sensitivity is neglected.

3.22. In Problem 3.17, the delta rosette has a factor of transverse sensitivity +2.3%. Determine the principal stresses and the principal angles. How much error will be introduced in determining the maximum principal stress if the effect of transverse strains is neglected?

3.23. Repeat Problem 3.21 if $K_t = -3.1\%$.

3.24. Repeat Problem 3.22 if $K_t = -2.3\%$.

REFERENCES

[1] C. Wheatstone, An account of several new instruments and processes for determining the constants of a voltaic circuit, *Philos. Trans. R. Soc.(London)*, **133**, 303 ff., 1843.

[2] W. M. Murray and P. K. Stein, *Strain Gage Techniques,* 1958.

[3] *Wheatstone Bridge Nonlinearity,* Tech. Note TN-507, Measurements Group, Inc., Raleigh (North Carolina), 1982.

[4] *High-Elongation Measurements,* Tech. Tip TT-605, Measurements Group, Inc., Raleigh (North Carolina), 1983.

[5] *Transverse Sensitivity Errors,* Tech. Note TN-509, Measurements Group, Inc., Raleigh (North Carolina), 1982.

[6] A. S. Khan and M. Malik, *Elementary Solid Mechanics,* in review.

[7] *Shunt Calibration,* Tech. Note TN-514, Measurements Group, Inc., Raleigh (North Carolina), 1988.

[8] *P-3500 Digital Strain Indicator,* Instr. Manual, Measurements Group, Inc., Raleigh (North Carolina), 1983.

[9] S. Huang and A. S. Khan, On the use of electrical-resistance metallic foil strain gages for measuring large dynamic plastic deformation up to forty percent, *Exp. Mech.*, vol. 31, pp. 122–125, 1991.

C H A P T E R   4

Photoelasticity

4.1 INTRODUCTION

Photoelasticity is one of the major experimental techniques for analyzing stress or strain distributions in loaded members. *Photo* depicts the use of optical methods, and *elasticity* implies interpretation of the experimental data utilizing the theory of elasticity. However, by using models made of transparent polymers, or through the technique of photoelastic coating, the applications have been extended to inelastically deforming bodies [3].

With the development of computers, the traditional areas of application of photoelasticity have been partially replaced by modern numerical techniques. However, developments in lasers, fiber optics, image analysis, and data acquisition continue to extend the range of applications, for example in fracture mechanics, so that the use of the photoelastic technique is still increasing instead of decreasing in stress or strain analysis [2]. This chapter is intended only to introduce the basic principles and techniques. The reader should refer to Dally and Riley [1], a chapter on photoelasticity by Burger in [2], Kuske and Robertson [6], Frocht [13,14], Cloud [16], and the selected scientific papers of M. M. Frocht [5] for more information.

4.2 REVIEW OF THE NATURE OF LIGHT AND WAVE REPRESENTATION

Maxwell showed that the behavior of light could be explained by assuming that light was electromagnetic radiation. The electromagnetic radiation, according to the simple wave theory, consists of transverse waves of very short wavelength which are propagating at an extremely high velocity (about 186,000 miles per second or 3×10^8 m/s in a vacuum or in air). The vector representing the light wave, called the light vector, can be either the electric vector or the magnetic vector. However, experiments show that the electric vector is the active agent in interactions between light and a photographic plate, so all subsequent discussions about the light wave will be as the electric vector.

For simplicity and convenience of representation, only a single harmonic wave of light is considered herein, since any other complicated waveform, such as the sawtooth waveform or the square waveform, can be represented for mathematical analysis by a Fourier series. Accordingly, light propagating in the positive x direction away from the source at a velocity v can be expressed as

$$L = A_L \cos\left[\frac{2\pi}{\lambda}(x - vt)\right]$$

(4.1a)

where L is the magnitude of the light vector, A_L the amplitude, λ the wavelength, z the position along the axis of propagation, v the velocity of propagation, and t the time.

Using the well-known Euler's formula, $e^{jx} = \cos x + j \sin x$, Eq. (4.1a) can conveniently be expressed as

$$L = \text{Re}[A_L e^{j[(2\pi/\lambda)(x-vt)]}]$$

(4.1b)

where Re is the real part of the complex variable and $j = \sqrt{-1}$.

The time required for the wave to travel a distance of one wavelength is defined as the period T, which is given by

$$T = \frac{\lambda}{v}$$

(4.2)

The inverse of T, or f, is called the *frequency*, which is the number of oscillations per second. Thus

$$f = \frac{1}{T} = \frac{v}{\lambda}$$

(4.3)

or

$$v = \lambda f$$

(4.4)

The angular frequency ω is given by

$$\omega = \frac{2\pi}{T} = 2\pi f$$

(4.5)

Therefore, Eq. (4.1b) can be rewritten as

$$L = \text{Re}[A_L e^{j(2\pi x/\lambda - \omega t)}]$$

(4.6)

In general, the wave may have an initial phase δ. Also Re is customarily omitted for simplicity. Thus Eq. (4.1b) becomes

$$L = A_L e^{j[(2\pi/\lambda)(x+\Delta-vt)]}$$

(4.7)

The range of the spectrum of visible light is a narrow band, ranging approximately 15.7 to 27.5 μin. (400 to 700 nm) in wavelength. Within this range, different wavelengths are interpreted as different colors by the human eye. The wavelengths of typical colors are given in Table 4.1. In other words, the color the eye observes depends on the frequency or combination of frequencies, since the velocity v is a constant [Eq. (4.4)].

TABLE 4.1   Spectrum of Visible Lights

Color	Wavelength [μin. (nm)]	Frequency (10^{14} Hz)
Violet	16.5 (420)	7.14
Blue	18.5 (470)	6.38
Blue-green	19.7 (500)	6.0
Green	20.9 (530)	5.66
Yellow (sodium)	23.2 (589.3)	5.09
Orange	23.6 (600)	5.0
Red	25.9 (660)	4.55

Source: Reference [7].

The effect of white light is produced when the complete spectrum is observed simultaneously. If the light is composed of a single wavelength, it is known as *monochromatic light.* The brightness or the intensity of light depends on the amplitude of the waves, which is proportional to the square of the amplitude, namely,

$$I = C'LL^* = CA_L^2 \tag{4.8}$$

where I is the intensity of the light, $C(C')$ is a constant, A_L is the amplitude, and L^* is the complex conjugate of L.

Consider two waves, waves 1 and 2, which have the same amplitude and wavelength but a different phase and are given by

$$L_1 = A_L e^{j[(2\pi/\lambda)(x+\Delta_1-vt)]}$$
$$L_2 = A_L e^{j[(2\pi/\lambda)(x+\Delta_2-vt)]} \tag{4.9}$$

The linear phase difference, $\Delta = \Delta_2 - \Delta_1$, is defined as the difference between the x coordinates of the two waves at any one instant and is also known as *retardation* since wave 2 follows wave 1 if $\Delta_2 > \Delta_1$, as shown in Fig. 4.1.

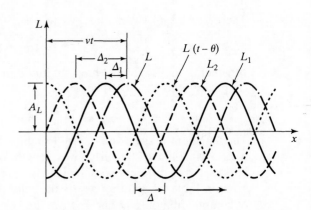

FIGURE 4.1   Retardation between two harmonic waves.

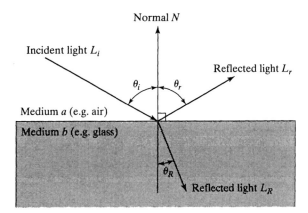

FIGURE 4.2 Reflection and refraction of light striking a surface.

The velocity in any other medium is smaller than the velocity in a vacuum. The ratio of the velocity in a vacuum to the velocity in a medium b is known as the *index of refraction i_b*, which is a property of the medium and will always be greater than unity. Thus

$$i_b = \frac{\text{velocity of light in vacuum } (v_0)}{\text{velocity of light in medium } b \ (v_b)} \qquad (4.10)$$

where v_0 is approximately 3×10^8 m/s or 11.8×10^9 in/s. Since the frequency of the light wave does not depend on the medium being traveled, the index of refraction can also be expressed in terms of the ratio of wavelengths, namely,

$$i_b = \frac{v_0}{v_b} = \frac{\lambda_0}{\lambda_b} \qquad (4.11)$$

Obviously, the wavelength is shorter in a medium than in a vacuum since the index of refraction is greater than unity.

When a beam of light is incident on a surface between two transparent mediums a and b (e.g., air and glass), which are optically isotropic and have different indices of refraction i_a and i_b, experiments show that the light is generally divided into two rays, a reflected ray and a refracted ray. Both rays lie in the plane formed by the incident ray and the normal to the surface, as shown in Fig. 4.2. In the figure, angles θ_i, θ_r, and θ_R are the angle of incidence, angle of reflection, and angle of refraction, respectively. The laws of reflection and refraction state that

$$\theta_i = \theta_r$$
$$\frac{\sin \theta_i}{\sin \theta_R} = \frac{i_b}{i_a} = i_{ba} \qquad (4.12)$$

where i_a is the index of refraction of medium a, i_b is the index of refraction of medium b, and i_{ba} is the index of refraction of medium b with respect to medium a.

EXAMPLE 4.1

Derive an expression for the linear phase shift δ for a light wave propagating in a material with respect to a similar light wave propagating in a vacuum. Assume that the thickness of the material along the propagating path of the light wave is d.

Solution   The time required for passage through a material of thickness d with velocity v_b is given by

$$t = \frac{d}{v_b}$$

The distance x traveled during the same time by a similar wave in a vacuum is

$$x = v_0 t = \frac{v_0 d}{v_b} = i_b d$$

where i_b is the index of refraction defined by Eq. (4.10), namely,

$$i_b = \frac{v_0}{v_b} \tag{a}$$

Therefore, the linear phase shift Δ is given by

$$\Delta = x - d = i_b d - d = d(i_b - 1) \tag{b}$$

4.3   POLARIZATION OF LIGHT

An unpolarized light beam is made up of a large number of transverse waves that are randomly oriented. Accordingly, a polarized light beam is made up of a number of transverse waves that have a preferred orientation. There are three polarized lights: elliptically polarized light, circularly polarized light, and plane or linearly polarized light, as shown in Fig. 4.3.

Elliptically polarized light, shown in Fig. 4.3b, is produced when the tips of the light vector along the z axis form an elliptical helix (either right- or left-handed helix). For example, the vectorial sum of two component waves having the same frequency but mutually perpendicular planes of vibration forms an elliptically polarized light. *Circularly polarized light* is a special case of elliptically polarized light obtained when the tips of the light vector along the z axis form a circular helix. The projections of these two lights on the x–y plane are shown in Fig. 4.3a. Another special case of elliptically polarized light occurs when the two component waves are in phase or the linear phase difference δ between the two components waves is zero. For this case, the light is known as *plane* or *linearly polarized light,* which exhibits a preferential plane of vibration, as shown in Fig. 4.3c.

In practice, there are several ways to polarize an ordinary light from a natural source. Polarized light can be obtained by reflection or refraction, by scattering, by fine grids, by using Polaroid sheets, and by birefringence (or double refraction)[2]. For most modern polariscopes, the Polaroid sheet is an extremely effective polarizing medium. The two important optical elements for experimental stress analysis by

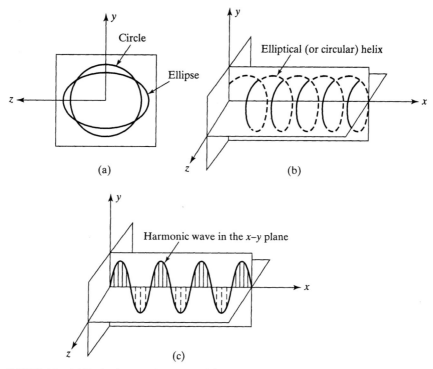

FIGURE 4.3   (a) Projection on plane of y-z (b) elliptically, or circularly, and (c) plane or linearly polarized light.

photoelasticity, the wave plate and the linear polarizer, are discussed in the following two sections.

4.4   WAVE PLATE AND THE PRINCIPLE OF PHOTOELASTICITY

The *wave plate,* also known as a *retarder* or *phase shifter,* is an optical element, which can resolve a light vector into two mutually perpendicular component waves and transmit both waves at different velocities, as shown in Fig. 4.4. Such a material with two different indices of refraction in the two perpendicular directions is known as a *birefringent* or *doubly refracting material.* In Fig. 4.4, axis f is referred to as the *fast axis* and axis s is referred to as the *slow axis,* since v_f is assumed to be greater than v_s. The angle θ_p is defined as the angle formed by the light vector and the fast axis. For a quarter-wave plate, which will be defined shortly, θ_p is set to $\pi/4$ in order to produce a circularly polarized light.

Since there is a difference in velocities, the components L_f and L_s will emerge from the birefringent material at different times, or there will be a phase difference between these two component waves. Consider first the component L_f, which is traveling at a velocity v_f. By Eq. (b) derived in Example 4.1, the distance Δ_f, by which the component wave L_f in the plate trails the wave in a vacuum, is given by

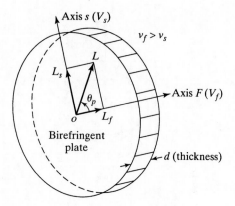

FIGURE 4.4   Wave plate.

$$\Delta_f = d(i_f - 1) \tag{c}$$

where $i_f = v_0/v_f$ is the index of refraction along the fast axis direction. Similarly, the distance Δ_s, by which the component wave L_s in the plate trails the wave in a vacuum, is given by

$$\Delta_s = d(i_s - 1) \tag{d}$$

Thus the relative linear phase shift can be expressed as

$$\Delta = \Delta_s - \Delta_f = d(i_s - i_f) \tag{4.13}$$

This serves as the basis for designing optical retarders. The relative angular phase shift Δ between these two components emerging from the plate can be proved to be

$$\delta = \frac{2\pi\Delta}{\lambda} = \frac{2\pi d(i_s - i_f)}{\lambda} \tag{4.14}$$

since the frequencies of the two component waves are the same.

Obviously, the relative angular phase shift produced by a birefringent plate depends on the thickness d, wavelength λ, and property of the plate represented by the term $(i_s - i_f)$. By adjusting these three factors, various values of δ can be obtained. If the plate is designed so that $\delta = \pi/2$, π, and 2π, it is known as a *quarter-wave plate*, a *half-wave plate*, and a *full-wave plate*, respectively.

EXAMPLE 4.2

Determine the relative angular phase shift (or retardation) δ in a quarter-wave plate employed with sodium light ($\lambda = 23.2$ μin.). The quarter-wave plate was designed originally for operation at $\lambda = 18.86$ μin.

Solution   At $\lambda = 18.86$ μin., $\delta = \pi/2$ (quarter-wave plate). From Eq. (4.14) we obtain

$$\frac{\pi}{2} = \frac{2\pi\Delta}{18.86} \quad \text{or} \quad \Delta = 4.72 \text{ μin.}$$

When the same quarter-wave plate operates at $\lambda = 23.2$ μin., the relative angular phase shift can be obtained using Eq. (4.14), namely,

$$\delta = \frac{2\pi\Delta}{\lambda} = \frac{(2\pi)(4.72)}{23.2} = 0.407\pi$$

It may be observed that certain noncrystalline transparent materials, which are optically isotropic under normal conditions, behave like wave plates when loaded (i.e., they temporarily become birefringent and have the ability to resolve a light vector into two orthogonal components transmitted with a different velocity). This phenomenon disappears when the applied load is removed, so it is called *temporary birefringent.* This physical characteristic is the basis of photoelasticity.

In a material exhibiting temporary birefringence, the changes in the indices of refraction are linearly proportional to the applied load or to the stresses if the material is linearly elastic. This observation was first reported by Maxwell [8] in 1853 and is known as the *stress-optic law.* The stress-optic law can be expressed in terms of relative retardation as follows:

$$\begin{aligned} i_1 - i_2 &= k(\sigma_1 - \sigma_2) \\ i_2 - i_3 &= k(\sigma_2 - \sigma_3) \\ i_3 - i_1 &= k(\sigma_3 - \sigma_1) \end{aligned} \qquad (4.15)$$

where $i_1, i_2,$ and i_3 are principal indices of refraction for waves vibrating parallel to the directions (1, 2, and 3) of principal stresses; k is the relative stress-optic coefficient, usually assumed to be a material constant (it actually varies with the wavelength); and $\sigma_1, \sigma_2,$ and σ_3 are the principal stresses at a point, respectively.

Consider a two-dimensional case, where the stressed model is a plate with its normal parallel to the direction of the propagation of the light. Assume that the in-plane principal stresses σ_1 and σ_2 do not vary through the thickness of the plate and σ_3 is zero (plane stress case). Since the stressed model behaves like a wave or retardation plate, the light vector will be resolved into two perpendicular components emerging from the plate with a relative retardation. By using Eqs. (4.13) and (4.15), we obtain

$$\Delta = dk(\sigma_1 - \sigma_2) \qquad (4.16)$$

where d is the thickness of the stressed plate. The relative angular retardation, Δ, is given by

$$\delta = \frac{2\pi\Delta}{\lambda} = \frac{2\pi dk(\sigma_1 - \sigma_2)}{\lambda} \qquad (4.17)$$

Equation (4.17) is the expression of the stress-optic law for a plane stress case. In engineering practice, the stress-optic law is more commonly expressed as

$$\sigma_1 - \sigma_2 = \frac{NM_f}{d} \qquad (4.18)$$

where $N = \delta/2\pi$ is the relative retardation in terms of a complete cycle of retardation and is also called the *isochromatic fringe order.* $M_f^g = \lambda/k$ (lb/in. or N/m) is known as the *material fringe value,* which is a material property for a given wavelength of light and is usually calibrated at the time of the test. Calibration methods are described in Section 4.8.

The stress difference $\sigma_1 - \sigma_2$ in a two-dimensional model can therefore be determined if the material fringe value M_f^{σ} can be determined by calibration and if the isochromatic fringe order can be measured experimentally at each point. This is actually achieved using a polariscope.

EXAMPLE 4.3

The material fringe value M_f^{σ} varies with the given wavelength of a monochromatic light. Assume that the material fringe value M_f^{σ} for a birefringent material was 36 lb/in. when a light with 18.86 μin. wavelength was employed during the determination of M_f^{σ}. Compute the material fringe value M_f for the same material if a light with 21.54-μin. wavelength is to be used.

Solution   Since $M_f^{\sigma} = \lambda/k$ or $k = \lambda/M_f^{\sigma} =$ material constant, thus

$$\frac{\lambda_1}{M_{f1}^{\sigma}} = \frac{\lambda_2}{M_{f2}^{\sigma}} = k$$

$$M_{f2}^{\sigma} = \frac{\lambda_2}{\lambda_1} M_{f1}^{\sigma} = \left(\frac{21.54}{18.86}\right)(36) = 41.1 \text{ lb/in.}$$

4.5   ANALYSIS OF A STRESSED PLATE IN A PLANE POLARISCOPE

A plane polariscope is an optical instrument that can produce a plane-polarized light and can measure the resulting phase difference when the polarized light passes through a stressed photoelastic model. The plane polariscope consists of a light source and two linear polarizers. The linear or plane polarizer is an optical element, which can resolve a light vector into two orthogonal components: one is transmitted and the other is absorbed, as shown in Fig. 4.5. The axis parallel with the transmitted component is called the *axis of polarization.* The H-type Polaroid sheet made from polyvinyl, for example, is the most familiar plane polarizer employed in modern polariscopy.

For illustration, let the light vector striking a plane polarizer be expressed as

$$L = A_L' e^{j\omega t} \tag{4.19}$$

where the initial phase is taken as zero for simplicity, since it is unimportant in photoelasticity. When the light enters the plane polarizer, it is resolved into two orthogonal components, one (L_t) parallel to the axis of polarization and transmitted; the other (L_a) perpendicular to the axis of polarization and absorbed. These components are given, respectively, by

$$L_t = A_L' \cos\theta_p e^{j\omega t} = A_L e^{j\omega t} \qquad L_a = A_L' \sin\theta_p e^{j\omega t} = = A_L^* e^{j\omega t} \tag{4.20}$$

where θ_p is the angle between the axis of polarization and the incident light vector, as shown in Fig. 4.5.

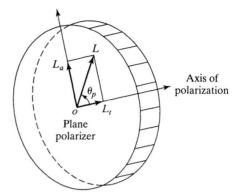

FIGURE 4.5 Linear or plane polarizer.

A commonly used plane polariscope is shown in Fig. 4.6. The plane polarizer near the light source is called the *polarizer* and its axis of polarization is called the *axis of the polarizer*. The other one is known as the *analyzer* and its axis of polarization is called the *axis of the analyzer*. Since the two axes are perpendicular to each other, no light will be transmitted through the analyzer when the transparent model is stress-free, and a dark field will result. For example, when the plate model, shown in Fig. 4.6, is subjected to an in-plane loading ($\sigma_3 = 0$), the principal stress σ_1 makes an angle α with the axis of polarizer. As described earlier, the polarizer resolves an incident light vector

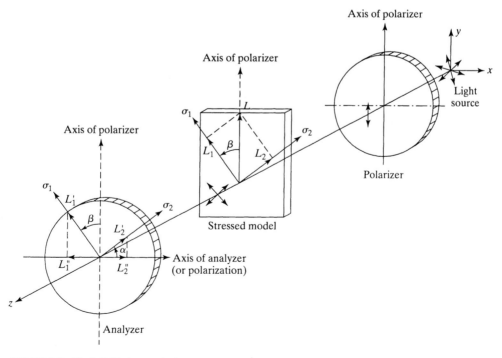

FIGURE 4.6 Dark-field plane polariscope.

into two orthogonal components, but only the one parallel to the axis of polarization will be transmitted. Using Eq. (4. 20a), the plane-polarized light vector L emerging from the polarizer will be

$$L = A_L e^{j\omega t} \tag{4.21}$$

This light vector enters the stressed plate, which behaves like a wave plate. The vector is decomposed into two orthogonal components in the directions of principal stress and are given by

$$
\begin{aligned}
L_1 &= A_L \cos \beta \; e^{j\omega t} \\
L_2 &= A_L \sin \beta \; e^{j\omega t}
\end{aligned}
\tag{4.22}
$$

which are shown in Fig. 4.6.

Let the thickness of the plate be d. Then the two component waves will travel with different velocities v_1 and v_2 through the thickness and emerge as

$$
\begin{aligned}
L_1' &= A_L \cos \beta \; e^{j(\omega t - \delta_1)} \\
L_2' &= A_L \sin \beta \; e^{j(\omega t - \delta_2)}
\end{aligned}
\tag{4.23}
$$

with

$$
\begin{aligned}
\delta_1 &= \frac{2\pi d}{\lambda} (i_1 - 1) \\
\delta_2 &= \frac{2\pi d}{\lambda} (i_2 - 1)
\end{aligned}
\tag{4.24}
$$

where δ_1 and δ_2 are the phase shift with respect to a wave in the air (note that the velocity in a vacuum is about 1.0003 times the velocity in the air). The two components will then enter the analyzer, but only the components of the two waves that are parallel to the axis of the analyzer will be transmitted, as shown in Fig. 4.6. Consequently, the final transmitted light vector L_a is given by

$$L_a = L_2'' - L_1'' = L_2' \cos \beta - L_1' \sin \beta \tag{4.25}$$

Substitution of Eq. (4.23) into Eq. (4.25) yields

$$
\begin{aligned}
L_a &= A_L \sin \beta \cos \beta \left[e^{j(\omega t - \delta_2)} - e^{j(\omega t - \delta_1)} \right] \\
&= A_L \sin 2\beta \sin \frac{\delta_2 - \delta_1}{2} e^{j[\omega t - (\delta_2 - \delta_1)/2 - \pi/2]}
\end{aligned}
\tag{4.26}
$$

Because the intensity of the light I is proportional to the square of the amplitude (the coefficient of the time-dependent term), Eq. (4.8) becomes

$$I = C A_L^2 \sin^2 2\beta \sin^2 \frac{\delta_2 - \delta_1}{2} = C A_L^2 \sin^2 2\beta \sin^2 \frac{\delta}{2} \tag{4.27}$$

where δ is the relative retardation and is given by

$$\delta = \delta_1 - \delta_2 = \frac{2\pi d(i_1 - i_2)}{\lambda} = \frac{2\pi dk(\sigma_1 - \sigma_2)}{\lambda} = \frac{2\pi d(\sigma_1 - \sigma_2)}{M_f} \tag{4.28}$$

Thus the average angular phase shift $(\delta_2 + \delta_1)/2$ has no influence on the intensity of the light, I, emerging from the analyzer, so it contributes nothing to the optical patterns (isoclinic or isochromatic fringe patterns) observed in a photoelastic model.

By examining Eq. (4.27), it is obvious that there are only two conditions for extinction of the light emerging from the analyzer.

1. The first case is when $\alpha = 0$ or $\pi/2$, $\sin^2 2\beta = 0$, so that extinction occurs. In other words, when either σ_1 or σ_2 is aligned with the axis of the polarizer, extinction will be produced irrespective of the particular wavelength of the light being used in the plane polariscope. Thus black fringe will be observed, which joins points or regions of the stressed plate where this condition is satisfied, since changes in stress directions occur in a continuous manner in the plate.

This fringe pattern, known as an *isoclinic fringe pattern,* is used to determine the directions of the principal stress at any point in a stressed model. For example, one can rotate the polarizer and the analyzer together, still with their axes of polarization perpendicular relative to the stressed model until an isoclinic fringe is coincident with a given point on the model. (Note that the isochromatic fringes, which will be defined below, will not be affected by the rotation.) The directions of the principal stress are determined immediately by inclination of the axes of the polarizer and the analyzer. Figure 4.7 shows a picture of an analyzer dial in a plane polariscope. The test point is covered by an isoclinic fringe by rotating the crossed polarizer and analyzer together. The principal stress (or strain) direction can be read directly from the calibrated dial.

2. Extinction will also occur when $\delta/2 = n\pi$, where $n = 0, 1, 2, \ldots$, since $\sin^2 \delta/2 = 0$. By noting Eq. (4.28), extinction will occur when the difference of principal stresses is either zero ($n = 0$) or an integral number of wavelengths ($n = 1,2,\ldots$). Since changes in stress also occur in a continuous manner in the model, fringes of various order will be observed. These fringes are known as *isochromatics.*

Examination of Eq. (4.28) indicates that the isochromatic fringe position depends on the wavelength (λ) of the light being used and the difference ($\sigma_1 - \sigma_2$) of the principal stresses. It is independent of the relative position (α) of the perpendicular polarizer and analyzer. When monochromatic light (single wavelength) is used, the

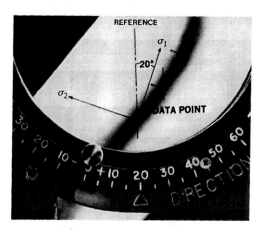

FIGURE 4.7   Dial on the Model 031 polariscope [11].

isochromatic fringe pattern appears as a series of black bands since the intensity of light is zero. If a white light (consisting of all wavelengths of the visible spectrum) is used, the isochromatic fringe patterns appear as a series of colored bands, since the difference of principal stresses in general produces extinction only for a particular wavelength. Only when $(\sigma_1 - \sigma_2) = 0$ (at isotropic points) or n = 0 will black isochromatic fringe be observed under white light.

In general, the isoclinic and the isochromatic fringe patterns are superimposed if the model is viewed in a plane polariscope. In practice, the plane polariscope with white light is frequently employed to obtain the isoclinic fringe (black fringes) patterns, so that the principal directions at any point in the model can be determined. It should be emphasized that not all the black fringes from plane-polarized light are necessarily isoclinics. Points of zero stress or points of equal principal stresses will also cause dark spots or black fringes. To eliminate the possibility of errors in sketching isoclinics, we should compare the fringe patterns with and without quarter-wave plates since the effect of quarter-wave plates, among other things, is to remove the isoclinics [5]; or we should rotate the crossed polarizer and analyzer together at a fixed load. Those black fringes that do not change their locations during the rotation are isochromatics.

4.6 ANALYSIS OF A STRESSED PLATE IN A CIRCULAR POLARISCOPE

The circular polariscope, which contains two more quarter-wave plates than those of the plane polariscope and produces circularly polarized light, is another optical instrument that can measure the phase difference occurring when circularly polarized light passes through a stressed photoelastic model, as shown in Fig. 4.8. The first quarter-wave plate converts the linearly polarized light emerging from the polarizer into circularly polarized light. The second quarter-wave plate converts the circularly polarized light back into linearly polarized light. The vibrating plane of the light depends on the relative position of the fast axes of the two wave plates. In Fig. 4.8, for example, the fast (or slow) axes of the two wave plates are perpendicular to each other, so the light will vibrate in the vertical plane and a dark field will result when the model is stress-free. There are four possible arrangements. The two most commonly used are listed in Table 4.2. Example 4.4 shows how quarter-wave plates with their fast (or slow) axes at right angles can minimize or cancel out any error introduced by imperfect quarter-wave plates.

TABLE 4.2   Two Commonly Used Arrangements in a Circular Polariscope

Arrangement	Axes of Polarizer and Analyzer	Fast (Slow) Axes of Quarter-Wave Plates	Field
1	Perpendicular	Perpendicular	Dark
2	Parallel	Perpendicular	Light

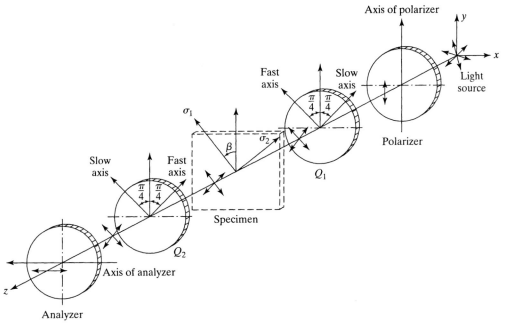

FIGURE 4.8   Dark-field circular polariscope [1,2].

EXAMPLE 4.4 [1]

Prove that the arrangement of two wave plates perpendicular to each other in a circular polariscope, as shown in Fig. E4.4, can cancel out a portion of the error introduced by the imperfection of the quarter-wave plates [i.e., if both quarter-wave plates differ from $\pi/2$ by a small amount ϵ, say, $\delta = (\pi/2) - \epsilon$].

Solution   1. Light from the polarizer can be expressed as $L = A_L e^{j\omega t}$.

2. Two light components along the fast and slow axes after entering the first wave plate can be written as

$$L_{F1} = A_L \cos \frac{\pi}{4} e^{j\omega t} = A'_L e^{j\omega t}$$

$$L_{S1} = A_L \sin \frac{\pi}{4} e^{j\omega t} = A'_L e^{j\omega t}$$

3. Light components after emerging from the first wave plate:

$$L_{F11} = A'_L e^{j\omega t}$$
$$L_{S11} = A'_L e^{j[\omega t - (\pi/2 - \epsilon)]}$$

since the phase shift is $(\pi/2) - \epsilon$ between these two components, due to the property of the quarter-wave plate.

4. Light components entering the second wave plate:

$$L_{F2} = L_{S11} = A'_L e^{j[\omega t - (\pi/2 - \epsilon)]}$$
$$L_{S2} = L_{F11} = A'_L e^{j\omega t}$$

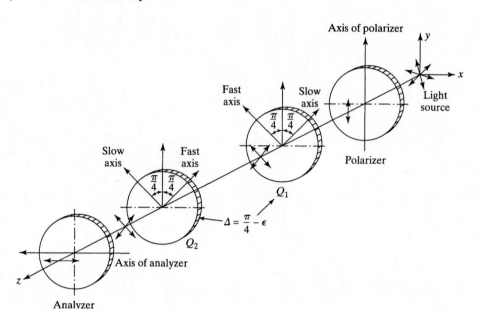

FIGURE E4.4

5. Light components after leaving the second wave plate:

$$L_{F22} = A'_L e^{j[\omega t - (\pi/2 - \epsilon)]}$$
$$L_{S22} = A'_L e^{j[\omega t - (\pi/2 - \epsilon)]}$$

since the relative phase shift is $\pi/2 - \epsilon$ between these two components according to the property of the quarter-wave plate.

Therefore, there is no phase shift between these two components that will enter the analyzer. A portion of the error that was introduced by imperfect quarter-wave plates is canceled. This is the reason why arrangements 1 and 2 in Table 4.2 are preferred in practice over dark- and light-field polariscopes where the axes of the two quarter-wave plates are parallel.

As mentioned in the preceding section, isoclinic and isochromatic fringe patterns are superimposed if the model is viewed in a plane polariscope. Therefore, the circular polariscope can be employed to eliminate the isoclinic fringe pattern and to exhibit the isochromatic fringe pattern only.

Let us analyze, step by step, the optical effects of a stressed model in a circular polariscope. Consider the stressed plate placed in the dark-field circular polariscope (arrangement 1), shown in Fig. 4.8. To simplify the following analysis, let the plane-polarized light emerging from the polarizer be

$$L = \sqrt{2} A_L e^{j\omega t} \tag{4.29}$$

On entering the quarter-wave plate, the light vector is resolved into two orthogonal components, L_f and L_s, which are given by

$$L_f = A_L e^{j\omega t} \qquad L_s = A_L e^{j\omega t} \tag{4.30}$$

where the subscripts f and s represent the fast axis and slow axis, respectively.

By the property of the quarter-wave plate, one component (along the slow axis) is retarded by $\pi/2$ relative to the other component (along the fast axis) after they emerge from the quarter-wave plate. Thus

$$L'_f = A_L e^{j\omega t} \qquad L'_s = A_L e^{j(\omega t - \pi/2)} \tag{4.31}$$

Equation (4.31) represents a circularly polarized light vector. On entering the stressed plate, which behaves like a wave plate, these two components are resolved into

$$
\begin{aligned}
L_1 &= L'_f \cos\left(\frac{\pi}{4} - \beta\right) + L'_s \sin\left(\frac{\pi}{4} - \beta\right) \\
L_2 &= -L'_f \sin\left(\frac{\pi}{4} - \beta\right) + L'_s \cos\left(\frac{\pi}{4} - \beta\right)
\end{aligned}
\tag{4.32}
$$

where L_1 and L_2 are components in the σ_1 and σ_2 directions, as shown in Fig. 4.9. Substitution of Eq. (4.31) into Eq. (4.32) yields

$$
\begin{aligned}
L_1 &= A_L \cos\left(\frac{\pi}{4} - \beta\right) e^{j\omega t} + A_L \sin\left(\frac{\pi}{4} - \beta\right) e^{j(\omega t - \pi/2)} \\
&= A_L e^{j[\omega t - (\pi/4 - \beta)]} \\
L_2 &= -A_L \sin\left(\frac{\pi}{4} - \beta\right) e^{j\omega t} + A_L \cos\left(\frac{\pi}{4} - \beta\right) e^{j(\omega t - \pi/2)} \\
&= A_L e^{j[\omega t - (\pi/4 - \beta) - \pi/2]}
\end{aligned}
\tag{4.33}
$$

Since only the relative phase difference is important, the common phase change $(\pi/4 - \beta)$ in Eq. (4.33) may be ignored to simplify the presentation:

$$
\begin{aligned}
L_1 &= A_{L1} e^{j\omega t} \\
L_2 &= A_{L1} e^{j(\omega t - \pi/2)}
\end{aligned}
\tag{4.34}
$$

After passing through the stressed plate with different velocities, the two components have a relative retardation Δ and become

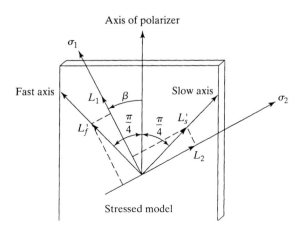

FIGURE 4.9   Decomposition of the light components entering the stressed plate.

$$L_1' = A_{L1}e^{j\omega t}$$
$$L_2' = A_{L1}e^{j(\omega t - \delta - \pi/2)}$$

(4.35)

where δ is given by Eq. (4.28).

On entering the second quarter-wave plate, whose fast and slow axes are interchanged in relation to the first quarter-wave plate (i.e., each axis of the two quarter-wave plates is perpendicular to the other), these two components (refer to Fig. 4.10) are resolved into

$$L_s = L_1' \cos\left(\frac{\pi}{4} - \beta\right) - L_2' \sin\left(\frac{\pi}{4} - \beta\right)$$
$$L_f = L_1' \sin\left(\frac{\pi}{4} - \beta\right) + L_2' \cos\left(\frac{\pi}{4} - \beta\right)$$

(4.36)

Substituting Eq. (4.35) into the equations above yields

$$L_s = A_{L1} \cos\left(\frac{\pi}{4} - \beta\right) e^{j\omega t} - A_{L1} \sin\left(\frac{\pi}{4} - \beta\right) e^{j(\omega t - \delta - \pi/2)}$$
$$L_f = A_{L1} \sin\left(\frac{\pi}{4} - \beta\right) e^{j\omega t} + A_{L1} \cos\left(\frac{\pi}{4} - \beta\right) e^{j(\omega t - \delta - \pi/2)}$$

(4.37)

On emerging from the second quarter-wave plate, the component L_s will be retarded by $\pi/2$ relative to component L_f. Assuming that L_f remains the same (since the common phase change has no influence on the analysis), we have

$$L_f' = L_f = A_{L1} \sin\left(\frac{\pi}{4} - \beta\right) e^{j\omega t} + A_{L1} \cos\left(\frac{\pi}{4} - \beta\right) e^{j(\omega t - \delta - \pi/2)}$$
$$L_s' = A_{L1} \cos\left(\frac{\pi}{4} - \beta\right) e^{j(\omega t - \pi/2)} - A_{L1} \sin\left(\frac{\pi}{4} - \beta\right) e^{j(\omega t - \delta - \pi)}$$
$$= A_{L1}\left[\cos\left(\frac{\pi}{4} - \beta\right) e^{j(\omega t - \pi/2)} + \sin\left(\frac{\pi}{4} - \beta\right) e^{j(\omega t - \delta)}\right]$$

(4.38)

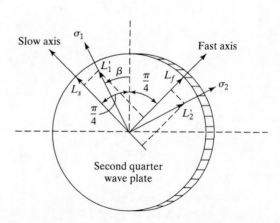

FIGURE 4.10 Decomposition of the light components entering the second wave plate.

These two components finally enter the analyzer, as shown in Fig. 4.11, but only their horizontal components (parallel to the axis of analyzer) will be transmitted. Thus

$$L_A = L_{sA} - L_{fA} = \frac{L'_s}{\sqrt{2}} - \frac{L'_f}{\sqrt{2}} = \frac{\sqrt{2}\,(L'_s - L'_f)}{2} \tag{f}$$

Substituting Eq. (4.38) into Eq. (f) gives

$$
\begin{aligned}
L_A &= \frac{A_{L1}\sqrt{2}}{2}\left[\cos\left(\frac{\pi}{4}-\beta\right)e^{j(\omega t - \pi/2)} + \sin\left(\frac{\pi}{4}-\beta\right)e^{j(\omega t - \delta)}\right.\\
&\quad \left. - \sin\left(\frac{\pi}{4}-\beta\right)e^{j\omega t} - \cos\left(\frac{\pi}{4}-\beta\right)e^{j(\omega t - \delta - \pi/2)}\right]\\
&= \sqrt{2}\,A_{L1}\sin\left(\frac{\delta}{2}\right)e^{j(\omega t + \beta - \delta/2 - \pi/4)}
\end{aligned}
\tag{4.39}
$$

Equation (4.39) is of the form $L_A = A_L e^{j(\omega t - \theta)}$, where A_L is the amplitude and θ is the phase angle. The transmitted intensity of the light emerging from the analyzer of a dark-field circular polariscope is given by

$$
\begin{aligned}
I &= c\sin^2\frac{\delta}{2}\\
&= c\sin^2\frac{\pi dk}{\lambda}(\sigma_1 - \sigma_2)
\end{aligned}
\tag{4.40}
$$

where $c = 2CA_{L1}^2$. Examining Eq. (4.40) indicates that the intensity of the light is a function only of the difference of the principal stresses, or the shear stress, and that extinction will occur when $\delta/2 = n\pi$, where $n = 0, 1, 2, \ldots$. In other words, only isochromatic fringe patterns will be observed under a circular polariscope. By definition of the fringe order $(N = (\delta/2)/\pi)$, integral fringe orders can be observed using a dark-field circular polariscope.

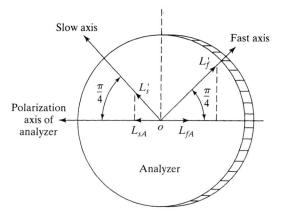

FIGURE 4.11   Decomposition of the components along the slow and fast axes of the second wave plate along the axis of analyzer giving the transmitted light.

In practice, the light-field (arrangement 2) circular polariscope is also frequently employed to obtain half-order fringes (i.e., $n = 1/2, 3/2, 5/2, \ldots$). This can be done simply by rotating the analyzer through $\pi/2$ from a dark-field circular polariscope arrangement. By using both dark- and light-field circular polariscopes, twice as many isochromatic fringes are obtained so that the accuracy is increased for determination of the shear stress $(\sigma_1 - \sigma_2)/2$ field for the whole model.

To establish the expression for the light emerging from the analyzer of a light-field circular polariscope, consider the components of the light emerging from the second quarter-wave plate, which are given by Eq. (4.38). On entering the analyzer shown in Fig. 4.12, only the vertical components (parallel to the axis of the analyzer) of these two components will be transmitted. Thus

$$L''_A = \frac{\sqrt{2}}{2}(L'_s + L'_f) \tag{g}$$

Substituting Eq. (4.38) into Eq. (g) gives

$$L''_A = \frac{A_{L1}\sqrt{2}}{2}\left[\cos\left(\frac{\pi}{4} - \beta\right)e^{j(\omega t - \pi/2)} + \sin\left(\frac{\pi}{4} - \beta\right)e^{j(\omega t - \delta)}\right.$$
$$\left. + \sin\left(\frac{\pi}{4} - \beta\right)e^{j\omega t} + \cos\left(\frac{\pi}{4} - \beta\right)e^{j(\omega t - \delta - \pi/2)}\right] \tag{4.41}$$
$$= \sqrt{2}\,A_{L1}\cos\left(\frac{\delta}{2}\right)e^{j(\omega t - \beta - \delta/2 - \pi/4)}$$

Equation (4.41) is also of the form $L''_A = A_L e^{j(\omega t - \theta)}$, where A_L is the amplitude and θ is the phase angle. Thus the transmitted intensity of the light emerging from the analyzer of a light-field circular polariscope (arrangement 2) is given by

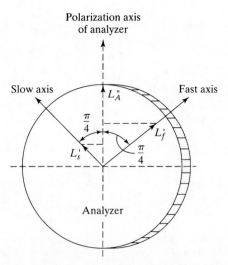

FIGURE 4.12   Transmission through the analyzer of the components along the slow and fast axes of the second wave plate (light field).

FIGURE 4.13 Isochromatic patterns for a ring subjected to a diametral compressive load: left, light field; right, dark field [9].

$$I'' = c \cos^2 \frac{\delta}{2} = c \cos^2 \frac{\pi dk}{\lambda} (\sigma_1 - \sigma_2) \qquad (4.42)$$

Examining Eq. (4.42) indicates that extinction will occur when $\delta/2 = n\pi + \pi/2$, where $n = 0, 1, 2, \ldots$. The fringe orders N will be $n + 1/2$. In other words, half-order fringes will be observed in a light-field circular polariscope (arrangement 2).

Figure 4.13 is an example showing the dark- and light-field isochromatic fringe patterns of a CR-39 ring subjected to a diametral compressive load; the left half shows half order fringes (i.e., $\frac{1}{2}, \frac{3}{2}, \frac{5}{2}, \ldots$) and the right half represents full order fringes (i.e., $1, 2, 3, \ldots$). By employing a circular polariscope with both dark- and light-field arrangements, a representation of isochromatic fringes is given to the nearest half order. By interpolation between fringes, the entire shear stress field can be determined accurately to $\pm 0.1 M_f^g/d$ [1,2]. If that accuracy is still not good enough, several special techniques can be employed to measure fractional fringe orders at selected points. In the next section we present one of the most commonly used techniques, the Tardy method.

4.7 TARDY METHOD

As described earlier, half-order fringes occur in a light-field circular polariscope arrangement, which is obtained simply by rotating the analyzer relative to the polarizer, in a dark-field circular polariscope through an angle $\pi/2$ (arrangement 1). A major advantage of this method is that no additional equipment is required. If the rotating angle is not $\pi/2$ but an arbitrary angle θ, fractional fringe orders at any point in the stressed plate can be determined. This method of analyzer rotation is known as the *Tardy method* [10].

Let us first establish the equation for light emerging from the analyzer rotated through an angle θ with respect to a dark-field circular polariscope (arrangement 1). Consider the components of the light emerging from the second quarter-wave plate, which are given by Eq. (4.38). On entering the analyzer shown in Fig. 4.14, only the

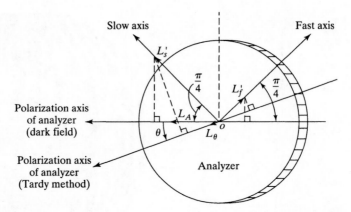

FIGURE 4.14 Transmitted components of the light emerging from the analyzer (Tardy method).

components parallel to the axis of the analyzer of these two waves will be transmitted. Thus

$$L_\theta = L'_s \cos\left(\frac{\pi}{4} + \theta\right) - L'_f \cos\left(\frac{\pi}{4} - \theta\right)$$

$$= \frac{\sqrt{2}}{2}\left[(L'_s - L'_f)\cos\theta - (L'_s + L'_f)\sin\theta\right] \qquad \text{(h)}$$

$$= L_A \cos\theta - L''_A \sin\theta.$$

Substituting Eqs. (4.39) and (4.41) into Eq. (h) and manipulating some trigonometric identities gives

$$L_\theta = \sqrt{2}\, A_{L1}\left(\sin\frac{\delta}{2}\cos\theta\, e^{j(\omega t + \beta - \delta/2 - \pi/4)} - \cos\frac{\delta}{2}\sin\theta\, e^{j(\omega t - \alpha - \delta/2 - \pi/4)}\right)$$

$$= \sqrt{2}\, A_{L1}\left\{\sin\frac{\delta}{2}\cos\theta\left(\cos\frac{\delta}{2}\, e^{j(\omega t + \beta - \pi/4)} + \sin\frac{\delta}{2}\, e^{j(\omega t + \beta - 3\pi/4)}\right)\right.$$

$$\left. - \cos\frac{\delta}{2}\sin\theta\left[\cos\left(2\beta + \frac{\delta}{2}\right)e^{j(\omega t + \beta - \pi/4)} + \sin\left(2\beta + \frac{\delta}{2}\right)e^{j(\omega t + \beta - 3\pi/4)}\right]\right\} \qquad (4.43)$$

$$= \sqrt{2}\, A_{L1}\left\{\left[\sin\frac{\delta}{2}\cos\frac{\delta}{2}\cos\theta - \cos\frac{\delta}{2}\sin\theta\cos\left(2\beta + \frac{\delta}{2}\right)\right]e^{j(\omega t + \beta - \pi/4)}\right.$$

$$\left. + \left[\sin\frac{\delta}{2}\sin\frac{\delta}{2}\cos\theta - \cos\frac{\delta}{2}\sin\theta\sin\left(2\beta + \frac{\delta}{2}\right)\right]e^{j(\omega t + \beta - 3\pi/4)}\right\}$$

$$= A_{L1}(1 - \cos\delta\cos 2\theta - \sin\delta\sin 2\theta\cos 2\beta)^{1/2}\, e^{j(\omega t + \beta - 3\pi/4 + \alpha)}$$

Equation (4.43) is of the form $L_\theta = A_{L1}e^{j(\omega t - \alpha')}$, where A_{L1} is the amplitude and α' is the phase angle. Thus the transmitted intensity of the light emerging from the analyzer is given by

$$I = C (1 - \cos \delta \cos 2\theta - \sin \delta \sin 2\theta \cos 2\beta) \qquad (4.44)$$

To have a zero value of light intensity (i.e., the transmitted intensity of the light reaches its minimum value since it is greater than or equal to zero), β must satisfy the equation

$$\frac{\partial I}{\partial \beta} = C(2 \sin \delta \sin 2\theta \sin 2\beta) = 0 \qquad (4.45)$$

Since in the Tardy method, θ is not equal to 0 or $\pi/2$ and δ is not equal to $n\pi$ ($n = 0,1,2,3, \ldots$), neither an integral-order nor half-order fringe will be observed. Therefore, β must take the value of either 0 or $\pi/2$. In other words, one of the directions of the principal stress must be parallel to the axis of the polarizer in the Tardy method so that extinction occurs ($I = 0$).

Substituting the values of β into Eq. (4.44) yields

$$I = C[1 - \cos(\delta \pm 2\theta)] \qquad (4.44a)$$

Hence the intensity of light transmitted by the analyzer will be zero when

$$\cos (\delta \pm 2\theta) = 1$$

or

$$\delta \pm 2\theta = 2n\pi \qquad (n = 0,1,2, \ldots)$$

By definition, the fringe order is given by

$$N = \frac{\delta/2}{\pi} = n \pm \frac{\theta}{\pi} \qquad (4.46)$$

To determine fractional fringe order at an arbitrary point A in a stressed model by the Tardy method, four major steps should ordinarily be followed.

1. The directions of principal stresses at point A must first be determined, since the direction of the principal stress must be parallel to the axis of polarizer in the Tardy method. A plane polariscope with white light is employed. Then one rotates the crossed polarizer and analyzer together until an isoclinic fringe is at point A. Now the axes of polarizer and analyzer are aligned with the directions of principal stresses σ_1 and σ_2 at point A.

2. Insert both quarter-wave plates with their axes crossed, and at $\pi/4$ to the axes of polarizer and analyzer, which results in a dark-field circular polariscope (arrangement 1). Thus the isoclinic fringes are removed and only the isochromatic fringes are retained. By inspection, the fringe order in the vicinity of a specific point can be determined. For example, if the point lies between fringes of orders n_1 and $n_1 + 1$, either n_1 or $n_1 + 1$ can be assigned to n in Eq. (4.46).

3. For illustration, let n_1 be assigned to n. Rotate the analyzer properly so that the fringe of order n_1 is moving toward point A. Continue rotating the analyzer through an angle θ until the fringe of order n_1 is coincident with point A. The fringe order at point A can then be determined by

$$N = n_1 + \frac{\theta}{\pi}$$

where a plus sign is used since the actual fringe order is greater than n_1.

4. To check for accuracy, we can rotate the analyzer in the opposite direction. The higher-order fringe $n_1 + 1$ will now be moving toward point A. Continue rotating the analyzer through an angle θ' until the fringe of order $n_1 + 1$ is coincident with point A. The fringe order at point A is then given by

$$N' = n_1 + 1 - \frac{\theta'}{\pi}$$

The minus sign is used since the actual fringe order is smaller than $n_1 + 1$. Theoretically, N is equal to N'.

It is clear that by employing the Tardy method, the fringe order at an arbitrary point within the stressed model can be effectively determined. No equipment other than a circular polariscope is required. The accuracy depends on the quality of the quarter-wave plates employed. Usually, the fringe order determined will be accurate to two decimal points [1]. Figure 4.15 shows a diagram of an analyzer dial used in practice for Tardy compensation. The dial is graduated in hundredths of a fringe, and θ/π can be read directly from the dial.

Finally, by using Eq. (4.18), the difference in principal stresses or the shear stress at point A can be obtained provided that the material fringe value M_f is known. How to determine the material fringe value M_f^σ experimentally, is described in the following section.

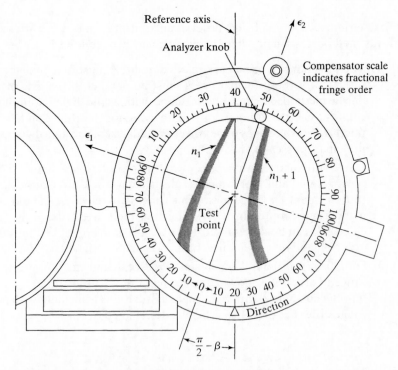

FIGURE 4.15 Analyzer dial used for Tardy Compensation [12].

4.8   CALIBRATION METHODS

To determine the stress distribution in a stressed model accurately, the material fringe value M_f^σ must be carefully calibrated. It is recommended that calibration be done for each sheet of photoelastic material at the time of the test, since even for the same photoelastic material used, the material fringe value M_f^σ can vary with the temperature, batch of resin, and age. Generally speaking, any model for which the theoretical stress distribution is accurately known can be employed to calibrate the material fringe value M_f^σ. The most common calibration loading systems are simple tension specimen, pure bending beam, and diametrically loaded disk [1, 2, 3].

Consider first a tensile specimen with thickness d and width w. The stresses at the test section are given by

$$\sigma_1 = \frac{F}{wd} \quad \text{and} \quad \sigma_2 = 0 = \sigma_3 \tag{i}$$

where F is the applied force. Substituting Eq. (i) into Eq. (4.18) yields

$$M_f = \frac{F}{Nw} \tag{4.47}$$

Figure 4.16 shows a typical plot of load F versus fringe order N. As the applied force is increased in a dark-field circular polariscope arrangement, the test section will be lighted and then become dark, giving the force for $N = 1$. This process will be repeated at $N = 2, 3, \ldots$. The linear regression method is used to draw the best straight line, which will not necessarily pass through zero since residual stresses in the model may be present. The slope of this line gives the value of F/N in Eq. (4.47).

A uniform beam subjected to four-point loadings, shown in Fig. 4.17a, is also commonly used to determine the material fringe value M_f^σ. The readings are taken in the portion where the beam is subjected to pure bending. For a given load P, it is easy to show that the principal stresses are given by

$$\sigma_1 = \sigma_y = \frac{12Pc}{dh^3} x \qquad \sigma_2 = \sigma_x = 0 = \sigma_3 \tag{j}$$

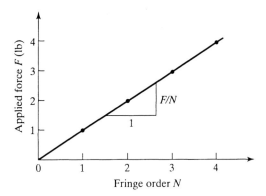

FIGURE 4.16   Force versus fringe-order diagram (tensile specimen).

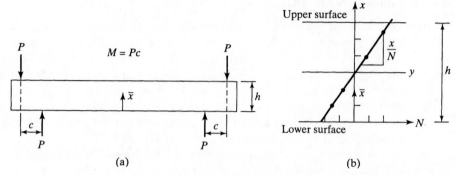

FIGURE 4.17   Four-point loaded beam and the distance versus fringe-order diagram.

where d is the thickness, h the height, and x the distance from the neutral axis y. Substituting Eq. (j) into Eq. (4.18) yields

$$M_f = \frac{12c}{h^3} \frac{x}{N} P \qquad (4.48)$$

As the load is increased, first-order fringes will appear at the top and bottom of the beam in the central constant bending moment region at a certain load. As the load increases, the first-order fringe will move toward the neutral axis. Subsequently, the second-order fringe will appear at the top and bottom, and several fringes will be seen above and below the neutral axis.

At a given load P, the distance x can be measured for each fringe-order location. A typical distance versus fringe-order diagram is shown in Fig. 4.17b. Note that the fringe orders below the neutral axis y are negative because σ_y is compressive and it must be designated σ_2. Again, the linear regression method is used to draw the best straight line. The intersection between this line and the x axis gives the position of the neutral axis y, namely, $x = 0$. Then the slope x/N can be obtained so that M_f^σ can be determined by Eq. (4.48). Actually, the location of the neutral axis is unimportant, since only the slope x/N is required to determine the material fringe value M_f^σ.

EXAMPLE 4.5

To determine the material fringe value M_f^σ, a uniform beam is subjected to four-point loading, as shown in Fig. 4.17a. The dimensions of the beam are $c = 0.75$ in., $h = 1.0$ in., and $d = 0.24$ in. At a load $P = 62.0$ lb, the isoclinic pattern is shown in Fig. E4.5a. Calculate the material fringe value M_f^σ.

Solution   From Fig. E4.5a we can obtain the following information:

$\bar{x}$ (in.)	0.19	0.31	0.43	0.54	0.66	0.78	0.91
N (fringe)	−3	−2	−1	0	1	2	3

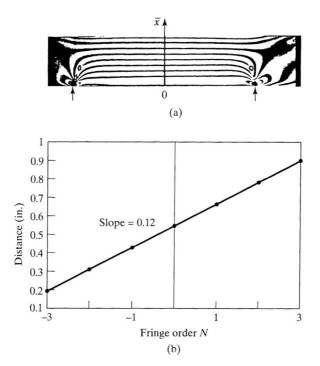

FIGURE E4.5

(b)

Thus a plot of distance x versus fringe order N is obtained, which is shown in Fig. E4.5b. The linear regression method is used to draw a straight line, and the slope (x/N) is approximately equal to 0.119. By using Eq. (4.48), the material fringe value M_f^σ can be calculated as

$$M_f^\sigma = (12)\,(0.75)\,(0.119)\,(62.0) = 66.4 \text{ lb/in.}$$

Note that the neutral axis of the beam is not at the middle due to the residual stresses in the model.

If the neutral axis is known exactly (as in the case of a beam having no residual stresses), another simple way can be employed to calculate M_f^σ. Consider a point A at the top or bottom of the beam (at $h/2$ distance away from the neutral axis); Eq. (4.48) now becomes

$$M_f^\sigma = \frac{6c}{h^2}\frac{P}{N} \tag{4.49}$$

Increase the applied load gradually until the first-order fringe appears at point A to obtain a point in the load versus fringe-order diagram shown in Fig. 4.16. Following the same procedure, several points can be obtained. Use the linear regression method to draw the best straight line and obtain the slope P/N so that M_f^σ can be determined by Eq. (4.49).

Calibration may also be achieved by using the circular disk model subjected to a diametral compressive load. Again, plot the load F against the fringe order N at the disk center. Then [1, 2, 12]

$$M_f^\sigma = \frac{8F}{\pi DN} \tag{4.50}$$

where D is the disk diameter. Equations (4.47) to (4.50) indicate that the value of M_f is independent of the model thickness d.

4.9   DETERMINATION OF WHOLE-FIELD ISOCLINIC FRINGE PATTERNS

The directions of the principal stresses can be determined if the isoclinic fringe patterns are known. Section 4.5 explains the procedure for determining the isoclinic fringe at any specified point in the model. The isoclinic fringe patterns over the entire model can be determined by combining a number of isoclinic patterns obtained at different plane polariscope settings (i.e., at different β values).

To illustrate this method, consider the example of a circular ring subjected to a diametral compressive load. Figure 4.18 shows six photographs of isoclinic fringe patterns taken at different settings (β is 0°, 15°, 30°, 45°, 60°, 75°). Under each picture is the corresponding sketch of the isoclinic fringe pattern. The combination of these six fringe patterns is shown in Fig. 4.19, where only half of the disk is shown due to symmetry.

For a plane stress case (plate model), the point at which $\sigma_1 = \sigma_2$ is known as an *isotropic point*. Furthermore, if $\sigma_1 = \sigma_2 = 0$, the point is called a *singular point*. It is obvious that all directions in the plane of the plate at either the isotropic point or the singular point are principal. For an optically isotropic material, isotropic points are easily identified by recognizing the fact that all isoclinics at different plane polariscope settings will pass through isotropic points. For example, point A (B or C) in Fig. 4.19 is an isotropic point. The isoclinic lines will rotate about an isotropic point when the crossed polarizers are rotated.

An isotropic point located at a free boundary (e.g., point B or C in Fig. 4.19) should also be a singular point. Such a point usually indicates a change of sign of the boundary stress. The point at which a concentrated load acts, functions as if it was an isotropic point. Isoclinics of all parameters will pass through that point, as shown in Fig. 4.18 or Fig. 4.19.

On a free boundary, or a boundary subjected to normal forces or pressure only (shear stress–free boundary), the directions of principal stresses are normal and tangential to the boundary. Thus an isoclinic line intersects the boundary at a point, for example, points D, E, F, G, H, I, and J in Fig. 4.19, where the inclination of the tangent is the same as the parameter of the isoclinic line. In other words, at any point on such a boundary, the acute angle between the tangent and the axis of polarization of the polarizer is the angle β (or $\pi/2 - \beta$) defined in Fig. 4.6.

The axis of symmetry (symmetric in both geometry and loading) always coincides with an isoclinic line, since the shear stress is zero on the symmetrical plane. In such a case, the isoclinic patterns are also symmetrical. However, an isoclinic line of parameter β on one side of the symmetry corresponds to an isoclinic line of the parameter $(\pi/2 - \beta)$ on the other side.

From the isoclinic patterns, however, one still cannot tell the individual directions of σ_1 and σ_2. This information may readily be gathered by observation of the fringe movement using the Tardy compensation technique. The experimental procedures are

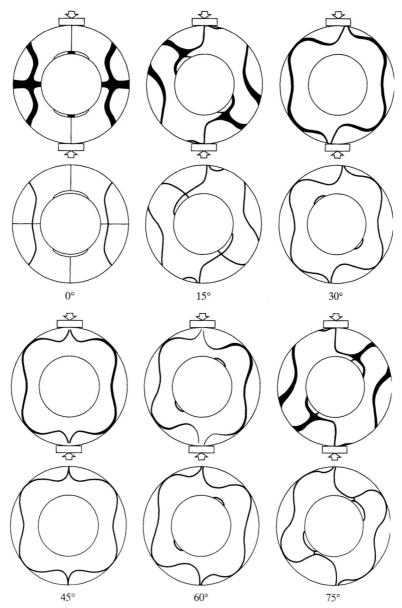

FIGURE 4.18   Isoclinic fringe patterns for a ring subjected to a diametral compressive load at different polariscope settings [12].

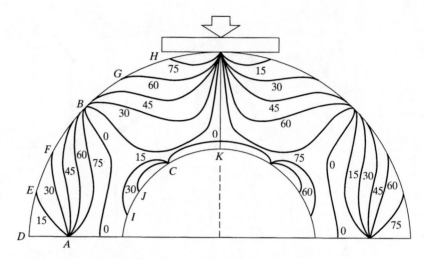

FIGURE 4.19   Sketch of isoclinics obtained by combinations [11].

as follows. First, calibrate the particular polariscope (say, a dark-field circular polariscope) being used with respect to a simple loaded specimen where the direction of σ_1 is known. For instance, the four-point loaded beam shown in Fig. 4.17a can be used in the calibration. The middle part of the beam is subjected to pure bending, so the direction of σ_1 is in the y direction.

With the axis of polarization of the analyzer parallel to the y axis (the direction of σ_1), the loaded beam model is viewed in a circular polariscope. Several fringes (say, fringes of order 0, 1, 2, and 3) will be in the field of view. Next, select a point in the beam model (say, a point between fringe orders 1 and 2), rotate only the analyzer (as for Tardy compensation) clockwise, and note whether the higher or lower fringe order moves toward that point.

Suppose that the lower-order fringe (order 1) moves toward the point in the beam model when rotating the analyzer clockwise. Then at any point in a stressed model, the direction of σ_1 will be determined by the same criterion; namely, if the lower-order fringe moves toward the point during clockwise rotation of the analyzer, the direction of σ_1 is aligned with the axis of polarization of the analyzer (before rotating). Conversely, if the higher-order fringe moves toward the point during clockwise rotation of the analyzer, the direction of σ_2 is aligned with the axis of polarization of the analyzer (before rotating). This is the so-called "calibration" of the polariscope.

Next, to determine the direction of σ_1 at any point in a stressed model, place the model in a plane polariscope and bring the isoclinic to the point by rotating the crossed polarizer and analyzer together. The axis of polarization of the analyzer will be parallel to either σ_1 or σ_2. Insert two quarter-wave plates to form a dark-field circular polariscope, which is the same as the one used in the calibration. Finally, rotate the analyzer clockwise and watch the movement of the fringes. If the lower-order fringe moves toward the point, the axis of polarization of the analyzer is aligned with σ_1. If the higher-order fringe moves toward the point, the axis of polarization of the analyzer (before rotation) is aligned with σ_2. In this way, the relative directions of σ_1 and σ_2 at any point in the stress model can quickly be determined.

The isoclinic parameters, together with isochromatic data, can also be used to determine the shear stress τ_{xy}. This value of shear stress τ_{xy} is then used to separate the principal stresses σ_1 and σ_2 by the shear-difference method described in the next section.

4.10   SEPARATION OF PRINCIPAL STRESSES

In the analysis of isochromatic fringe patterns, the fringe order N can be established by direct observation of the development of the fringes during the loading process. They will appear in the order $0, 1, 2, \ldots$ if a dark-field circular polariscope is used, or in the order $1/2, 3/2, 5/2, \ldots$ if a light-field circular polariscope is employed.

If fringe order development has not been observed directly and only photographs are at hand, however, determination of the fringe order is relatively more difficult. In general, the key step is to find or recognize the lowest fringe order. This can be achieved by using the approximate stress field. For instance, if there is a corner in the model that is stress-free, zero fringe order will appear at the corner; if the model is subjected to a combined tensile (or compressive) loading and bending moment, the zero-order fringe will appear at the neutral axis. Once the lowest-order fringe is identified, it is very easy to determine the fringe order at any point in the model by progressive counting of the fringe orders.

EXAMPLE 4.6

Fig. E4.6 shows the isochromatic fringe pattern for a fillet in bending. The picture is taken in a dark-field circular polariscope. Determine all the fringe orders.

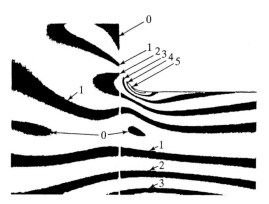

FIGURE E4.6   Isochromatic fringe pattern for a fillet in bending [11].

Solution   Since the fillet is stress-free at the top right corner in Fig. E4.6, the fringe is a zero-order fringe. It is a simple matter to determine the remaining fringe orders, which are orders of $1, 2, 3, 4,$ and 5, as also shown in Fig. E4.6.

Once the fringe order at one point in the model is known, the difference of principal stresses, $\sigma_1 - \sigma_2$, can be determined directly using Eq. (4.18). In practice, it is sometimes important to determine the complete state of stress at a point (i.e., σ_1, σ_2, and β). However, the combined isoclinic and isochromatic data provide only $(\sigma_1 - \sigma_2)$ and β. Thus additional information may be needed to separate the principal stresses σ_1 and σ_2.

Free-Boundary Stresses

Stress on the free boundary is important since the critical stress in a body frequently occurs at a boundary point. When the point is on a stress-free boundary, the principal stress normal to the boundary is zero; therefore, the remaining principal stress is determined directly by the isochromatic fringe order. The sign of the stress depends on whether the stress in the tangential direction is σ_1 or σ_2. That can be found readily either by the method described in Section 4.9 or by the much simpler method of applying a small normal force at the boundary point of interest and observing the change in the fringe order at that point. The tangential principal stress is σ_1 if the fringe order is increasing during application of the small load in the normal direction, and vice versa. If the stress is σ_1, it is positive since $\sigma_2 = 0$ and $\sigma_1 \geq \sigma_2$. Conversely, if the stress is σ_2, it is negative since $\sigma_1 = 0$ and $\sigma_1 \geq \sigma_2$. Figure 4.20a & b shows stresses at points on the free boundary.

At a free external corner, as shown in Fig. 4.20c, the complete absence of external forces must be accompanied by the total absence of stresses. Therefore, $\sigma_1 = \sigma_2 = 0$ at external corners on the free boundaries, so that the isochromatic fringe order is always zero. Knowing a fringe order in advance is extremely helpful in interpreting isochromatic patterns from a photograph. This is particularly true when the development of isochromatic fringe patterns during load application has not been observed directly (see Example 4.6).

Stress State at Interior Points

Several methods, both numerical and experimental, have been developed [1, 2, 15, 16] to provide the additional information necessary to determine the complete stress state at interior points. A few commonly used methods will be described below. These are

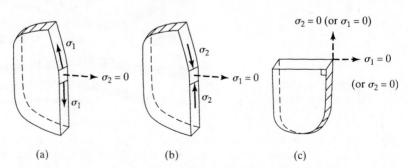

(a)　　　　　　　(b)　　　　　　　(c)

FIGURE 4.20　Free-boundary stresses [3].

based on Hooke's law, photoelasticity measurements at oblique incidence, equilibrium equations, and compatibility equations.

Method Based on Hooke's Law.   In a model subjected to a plane stress ($\sigma_z = 0$) loading condition, using Eq. (1.43), the lateral strain ϵ_z can be written as

$$\epsilon_z = \frac{-\mu(\sigma_x + \sigma_y)}{E} = \frac{-\mu(\sigma_1 + \sigma_2)}{E} \tag{4.51}$$

or

$$\sigma_1 + \sigma_2 = \frac{-E\epsilon_z}{\mu} = \frac{-E\,\Delta d}{\mu d} \tag{4.52}$$

where d is the thickness of the model, Δd the change of thickness, E is Young's modulus, and μ is Poisson's ratio. Thus it is possible to determine the sum of the principal stresses by measuring the change in thickness (Δd). Together with knowledge of the difference in the principal stresses determined by the isochromatic fringe order, the two principal stresses σ_1 and σ_2 can be separated. Obviously, the accuracy depends on measurement of the change in thickness. In practice, accurate lateral extensometers and various optical interferometers have been used to measure thickness changes.

Oblique-Incidence Method.   The stress-optic law, Eq. (4.18), was obtained under the condition that light was passing through the stressed model at normal incidence. In other words, the incident angle θ_i, shown in Fig. 4.2, is zero. However, if the stressed model is rotated so that the light passes through the model at an angle α, as shown in Fig. 4.21a, an oblique-incidence isochromatic fringe pattern can be obtained in a circular polariscope. This fringe pattern provides additional data, which can be used to separate the principal stresses σ_1 and σ_2.

Consider first the case where the direction of the principal stresses are known in advance, for instance, along the line of symmetry. Rotate the model about the σ_1 (or σ_2) axis (the relative directions of principal stresses can readily be determined by the method described in Section 4.9) clockwise by an angle α, as shown in Fig. 4.21a. Let

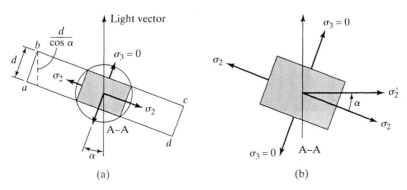

(a)   (b)

FIGURE 4.21   Rotation of the stressed model about the σ_1 axis by α.

the thickness of the model be d, so that light traverses a distance of $d/\cos \alpha$ through the oblique model. Now the fringe pattern developed in a circular polariscope is related to the principal stresses σ_1 and σ_2'. Therefore, Eq. (4.18) becomes

$$\sigma_1 - \sigma_2' = \frac{M_f^\sigma N_\alpha}{d/\cos \alpha} \tag{4.53}$$

where N_α is the fringe order obtained from the oblique-incidence method. Substituting $\sigma_{x'} \sigma_x, \sigma_y,$ and τ_{xy} with $\sigma_2, \sigma_2, 0,$ and $0,$ respectively (see Fig. 4.21b) in Eq. (I.7), we obtain

$$\sigma_2' = \sigma_2 \cos^2 \alpha \tag{4.54}$$

Thus

$$\sigma_1 - \sigma_2 \cos^2 \alpha = \frac{M_f^\sigma N_\alpha}{d/\cos \alpha} \tag{4.55}$$

Solving Eqs. (4.55) and (4.18) simultaneously yields

$$\sin^2 \alpha\, \sigma_1 = \frac{M_f^\sigma \cos \alpha}{d} (N_\alpha - N \cos \alpha)$$
$$\sin^2 \alpha\, \sigma_2 = \frac{M_f^\sigma}{d} (N_\alpha \cos \alpha - N) \tag{4.56}$$

where N is the fringe order obtained by the normal-incidence method. It is therefore possible to separate the principal stresses by using Eq. (4.56), provided that the isochromatic fringe patterns are known from the normal- and oblique-incidence methods. This method is often used in practice to separate the principal stresses along a line of symmetry by rotating the stressed model about the line (one of the principal stress axes). Figure 4.22 shows an example of isochromatic patterns of a circular disk in

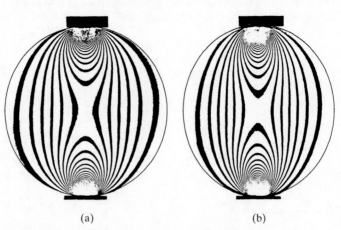

(a)   (b)

FIGURE 4.22   Isochromatic fringe patterns: (a) normal incidence; (b) oblique incidence (30° with the vertical axis) [14].

diametral compression for normal and oblique incidence with a 30° rotation about the vertical symmetry axis.

EXAMPLE 4.7

At a point in a stress model with a thickness of 0.24 in. and a material fringe value of 65 lb/in., a fringe order of 3.1 is obtained by the normal-incidence method and a fringe order of 4 is obtained by the oblique-incidence method, namely, by rotating the model about the σ_1 axis 30° in a clockwise direction. Determine the values of principal stresses.

Solution   Substituting $M_f^\sigma = 65$ in./lb, $d = 0.24$ in., $\alpha = 30°$, $N = 3.1$, and $N_\alpha = 4$ into Eq. (4.56), the principal stresses can readily be determined as

$$\sigma_1 = \frac{(65)\,(\cos 30°)}{(0.24)\,(\sin 30°)^2}\,(4 - 3.1\cos 30°) = 1234 \text{ psi}$$

$$\sigma_2 = \frac{65}{(0.24)\,(\sin 30°)^2}\,(4\cos 30° - 3.1) = 394.4 \text{ psi}$$

Consider now the general case where the directions of principal stresses are not known in advance. The principle is essentially the same as that in the case where the directions of principal stresses are known. Rotation about two arbitrary axes, say about the x and y axes, is necessary to provide an additional equation to match the additional unknowns.

Let the angles of rotation about the y axis be α. By Eqs. (4.53) and (1.12), we obtain

$$\sigma_{1'} - \sigma_{2'} = [(\sigma_{x'} - \sigma_{y'})^2 + 4\tau_{x'y'}^2]^{1/2} = \frac{M_f^\sigma N_{\alpha y}}{d/\cos\alpha} \tag{4.57}$$

where σ_1' and σ_2' are the principal stresses in the $x'y'$ plane, which is normal to incident light along the z' axis. $N_{\alpha y}$ is the fringe order obtained by the oblique-incident method associated with the model rotating about the y axis. Since

$$\sigma_{x'} = \sigma_x \cos^2\alpha \qquad \sigma_{y'} = \sigma_y \qquad \tau_{x'y'} = \tau_{xy}\cos\alpha$$

Eq. (4.57) becomes

$$\frac{M_f^\sigma \cos\alpha}{d}\,N_{\alpha y} = [(\sigma_x \cos^2\alpha - \sigma_y)^2 + 4t_{xy}^2 \cos^2\alpha]^{1/2} \tag{4.58}$$

Similarly, we obtain

$$\frac{M_f^\sigma \cos\alpha}{d}\,N_{\alpha x} = [(\sigma_x - \sigma_y \cos^2\alpha)^2 + 4\tau_{xy}^2 \cos^2\alpha]^{1/2} \tag{4.59}$$

where $N_{\alpha x}$ is the fringe order obtained by the oblique-incident method associated with the model rotation of α about the x axis. For simplicity, the angle of rotation about the x axis is assumed to be the same as the one about the y axis.

The normal incident pattern yields, by Eqs. (4.18) and (1.12),

$$\frac{M_f^\sigma}{d} N = (\sigma_1 - \sigma_2) = [(\sigma_x - \sigma_y)^2 + 4\tau_{xy}^2]^{1/2} \tag{4.60}$$

Solving Eqs. (4.58), (4.59), and (4.60) simultaneously for σ_x and σ_y gives

$$\sigma_x = \left(\frac{M_f^\sigma}{d}\right)\left\{\frac{\cot^2\alpha}{1 - \cos^4\alpha}\left[N_{\alpha x}^2 + N_{\alpha y}^2\cos^2\alpha - N^2(1 + \cos^2\alpha)\right]\right\}^{1/2}$$

$$\sigma_y = \left(\frac{M_f^\sigma}{d}\right)\left\{\frac{\cot^2\alpha}{1 - \cos^4\alpha}\left[N_{\alpha x}^2\cos^2\alpha + N_{\alpha y}^2 - N^2(1 + \cos^2\alpha)\right]\right\}^{1/2} \tag{4.61}$$

Using $\sigma_x + \sigma_y = \sigma_1 + \sigma_2$, together with Eq. (4.60), σ_1 and σ_2 can be separated.

EXAMPLE 4.8

Since the direction of principal stresses is not known, rotations about the x and y axes have been carried out. The angle is 30° for both axes. Derive expressions for σ_1 and σ_2 and the principal angle θ_1 in terms of M_f^σ, d, $N_{\alpha x}$, $N_{\alpha y}$, and N, whose definitions are apparent.

Solution   From Eq. (4.61) and $\alpha = 30°$, we obtain

$$\sigma_x = \frac{M_f^\sigma}{d}\left[\frac{12}{7}(4N_{\alpha x}^2 + 3N_{\alpha y}^2 - 7N^2)\right]^{1/2}$$

and

$$\sigma_y = \frac{M_f^\sigma}{d}\left[\frac{12}{7}(3N_{\alpha x}^2 + 4N_{\alpha y}^2 - 7N^2)\right]^{1/2}$$

Using Eq. (4.60), we have

$$\tau_{xy} = \frac{1}{2}\left[\frac{M_f^\sigma}{d}N - (\sigma_x - \sigma_y)^2\right]^{1/2}$$

Since

$$\sigma_1 + \sigma_2 = \sigma_x + \sigma_y$$

$$\sigma_1 - \sigma_2 = \frac{M_f^\sigma}{d}N$$

therefore

$$\sigma_1 = \frac{1}{2}\left(\sigma_x + \sigma_y + \frac{M_f^\sigma}{d}N\right)$$

$$\sigma_2 = \frac{1}{2}\left(\sigma_x + \sigma_y - \frac{M_f^\sigma}{d}N\right)$$

The principal angle can then be determined by Eq. (1.11b), namely,

$$\beta_1 \tan^{-1} \frac{\tau_{xy}}{\sigma_1 - \sigma_y}$$

Methods Based on the Equilibrium Equations

Shear-Difference Method.   The shear-difference method, based solely on the equations of equilibrium in Cartesian coordinates, is probably the most widely used method to separate stresses in a model along an arbitrarily chosen straight line drawn from a free boundary (or from a point where the stress state is known). In the case of plane stress with the absence of body forces, the equilibrium equations are given by Eq. (1.21). Thus the stress σ_x along the x axis can be obtained using the equation

$$(\sigma_x)_i = (\sigma_x)_{i-1} - \int \frac{\partial \tau_{xy}}{\partial y} \, dx \tag{4.62}$$

which can be approximated by the following finite-difference expression:

$$(\sigma_x)_i = (\sigma_x)_{i-1} - (\Delta\tau_{xy})_i^* \left(\frac{\Delta x}{\Delta y} \right)_i, \qquad i = 1,2,3,\dots \tag{4.63}$$

where

$$(\Delta\tau_{xy})_i^* = \frac{(\Delta\tau_{xy})_{i-1} + (\Delta\tau_{xy})_i}{2}$$

The term $(\sigma_x)_0$, i.e., $i = 1$, denotes the known stress at the starting point, which is generally located at the free boundary. For example, if a line is drawn from a point P on a free boundary, which is inclined at an angle ϕ to the boundary shown in Fig. 4.23a, then $(\sigma_x)_0 = \sigma_T \cos^2\phi$, where σ_T is the tangential boundary stress, and if $\phi = 90°$, $(\sigma_x)_0 = 0$. Using Mohr's stress circle (Section 1.5) and Eq. (4.18), the shear stress τ_{xy} can be computed using the equation

$$\tau_{xy} = \frac{1}{2}(\sigma_1 - \sigma_2) \sin 2\beta = \frac{NM_f^\sigma}{2d} \sin 2\beta \tag{4.64}$$

where β is the angle that the x axis makes with the direction of σ_1. The direction of σ_1 can be determined by the isoclinic patterns, as described in Section 4.9.

Lines MN and PQ are drawn parallel to ox, usually located symmetrically above and below this axis and are separated at Δy, as shown in Fig. 4.23a. τ_{xy} can be determined, using Eq. (4.64), along lines MN and PQ, and the τ_{xy} versus distance x curves can be drawn, as shown in Fig. 4.23b. The shaded area represents the quantity $\Sigma \, \Delta\tau_{xy} \, \Delta x$ in Eq. (4.63). Note that Δy is a constant and Δx is usually not a constant but is

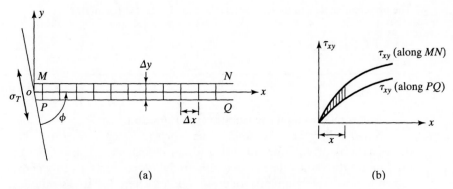

FIGURE 4.23   Grid system used in the shear-difference method.

dependent on the stress gradient. Thus σ_x can be obtained by graphical integration at every point along ox [Eq. (4.63)]. From the Mohr's stress circle, we have

$$\sigma_x = \frac{(\sigma_1 + \sigma_2) + (\sigma_1 - \sigma_2)\cos 2\beta}{2} \tag{4.65a}$$

or

$$\sigma_1 + \sigma_2 = 2\sigma_x - (\sigma_1 - \sigma_2)\cos 2\beta = 2\sigma_x - \frac{NM_f^{\sigma}\cos 2\beta}{d} \tag{4.65b}$$

Since $\sigma_1 + \sigma_2 = \sigma_x + \sigma_y$,

$$\sigma_y = \sigma_x - \frac{NM_f^{\sigma}\cos 2\beta}{d} \tag{4.66}$$

Knowing the sum $(\sigma_1 + \sigma_2)$ and difference $(\sigma_1 - \sigma_2)$ of the principal stresses, σ_1 and σ_2 along ox can be determined. This procedure can be repeated for any line of interest in the specimen provided that the stresses are known at the starting point.

EXAMPLE 4.9 [13]

Consider a plate with thickness 0.272 in. subjected to a tensile loading via a pin, as shown in Fig. E4.9a. The isoclinics are shown in Fig. E4.9b and the isochromatic fringe patterns (dark field) at a load of 247 lb are shown in Fig. E4.9c. Determine the stress distributions (σ_x, σ_y, and τ_{xy}) by the shear-difference method along line OH shown in Fig. E4.9a. The material fringe value is 85.68 lb/in.

Solution   1. Draw two auxiliary straight lines AB and CD, which are Δy (0.1w) apart, then draw the integration grid, as shown in Fig. E4.9a. For simplicity, the interval Δx equals Δy. In general, the interval is closer in the region where the stress gradient is greatest.

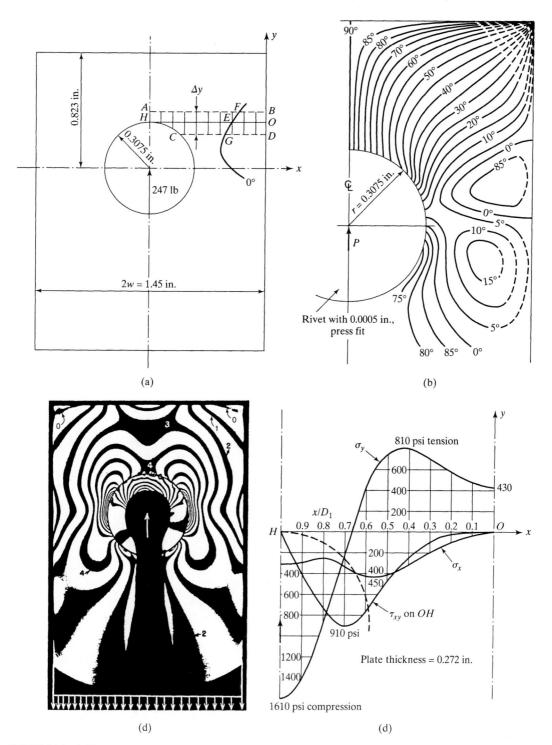

(a)

(b)

(d)

(d)

FIGURE E4.9   [12]

2. To find τ_{xy} by Eq. (4.64), it is very important to determine the angle θ properly from the isoclinics, since θ is the angle between the x axis and σ_1. The stresses at boundary points A, O, and C are tensile in the y direction. Thus $\sigma_1 = \sigma_y > 0$ and $\sigma_2 = \sigma_x = 0$, and θ should be 90°, which corresponds to the zero-order isoclinic fringe in Fig. E4.9a. At intervals FB, EO, and GD, σ_1 rotates clockwise from the y axis by a small angle; therefore, θ directly takes the isoclinic values shown in Fig. E4.9b. In other words, the isoclinic values for intervals FB, EO, and GD indicate the angle between σ_1 and the x axis. On the other hand, for intervals AF, HE, and CG, σ_1 rotates counterclockwise from the y axis, indicating that θ should take the values $\theta' + 90°$ (or $90° - \theta'$), where θ' has the isoclinic fringe orders shown in Fig. E4.9b. The values of θ along line OH at the grid points are given in the second column in Table E4.9.

3. The isochromatic fringe orders can be obtained readily by recognizing that the zero-order fringe is located at the top corners ($\sigma_x = \sigma_y = 0$), as shown in Fig. E4.9c. The isochromatic fringe orders N at grid points on line OH are given in the third column of Table E4.9. Once θ and N are known, τ_{xy} along line OH can be obtained using Eq. (4.64). These values appear in the fourth column of Table E4.9.

4. τ_{xy} at each grid point along lines BA and DC can be obtained in the same way. Then $\Delta\tau_{xy}$ along line OH can be calculated, namely, $\Delta\tau_{xy} = (\tau_{xy})_{BA} - (\tau_{xy})_{DC}$. See the fifth column in Table E4.9.

5. At the starting point o, $\sigma_x = 0$, and Eq. (4.63a) becomes

$$\sigma_x = -\Sigma \frac{\Delta\tau_{xy}}{\Delta y}\Delta x = \Sigma\,\Delta\tau_{xy}$$

since $\Delta x/\Delta y = -1$. The values of σ_x at the grid points are listed in the sixth column of Table E4.9.

6. Once σ_x along line OH is determined, σ_y can be calculated using Eq. (4.66). The seventh column of Table E4.9 shows these values.

7. The stress distributions are shown in Fig. E4.9d.

Filon's Method.   In some cases the isostatics are accurately known. For example, if the problem involves rotational symmetry, the isostatics are concentric circles

TABLE E4.9

Grid Point $(-x/w)$	θ (deg)	N	τ_{xy} Line OH (psi)	$\Delta\tau_{xy}$ Line OH (psi)	σ_x Line OH (psi)	σ_y Line OH (psi)
0	90.0	1.366	0.	0.	0.	430.3
0.1	88.25	1.714	16.5	−34.7	−34.7	504.2
0.2	91.0	2.222	−12.2	−70.9	−105.6	593.9
0.3	96.0	3.0	−98.2	−104.0	−209.6	714.7
0.4	102.5	4.032	−268.4	−121.3	−330.9	820.2
0.5	111.0	5.079	−535.3	−102.4	−433.3	755.6
0.6	121.0	5.810	−808.0	−12.0	−445.3	413.9
0.7	134.3	5.873	−924.7	111.8	−333.5	−288.3
0.8	148.5	5.159	−724.0	66.2	−267.3	−1005.3
0.9	164.0	4.349	−363.0	−37.8	−305.1	−1466.9
1.0	180.0	4.063	0.0	−17.3	−322.4	−1602.2

and radial lines. Therefore, it is convenient to use the Lamé–Maxwell equations, Eqs. (1.25) and (1.26). Integrating along one of the isostatics (say, along s_1) yields

$$\sigma_1 = (\sigma_1)_0 - \int \frac{\sigma_1 - \sigma_2}{\rho_2} \, ds_1 \qquad (4.67)$$

The integration procedure is similar to that described for the shear stress difference method. First, a starting point is selected at which $(\sigma_1)_0$ is known; the point is usually located on a free boundary. Plots containing the values of $(\sigma_1 - \sigma_2)/\rho_2$ versus s_1 are then obtained using the information of isoclinics and isochromatics. Thus the change in principal stress σ_1 between the starting point and the point of interest along the isostatic s_1 is the accumulated area under the curve in the diagram of $(\sigma_1 - \sigma_2)/\rho_2$ versus s_1.

This method is not widely used for general cases because it is difficult to obtain accurate data for the radius of curvature from the isoclinics by the photoelastic method. Therefore, detailed discussion of this method is omitted.

Methods Based on the Compatibility (Laplace's) Equation. As stated earlier, the compatibility equation for plane stress can be expressed in terms of stress components. In the Cartesian coordinate system, if body forces are absent, the *compatibility equation* is

$$\left(\frac{\partial^2}{\partial x^2} + \frac{\partial^2}{\partial y^2} \right)(\sigma_x + \sigma_y) = \left(\frac{\partial^2}{\partial x^2} + \frac{\partial^2}{\partial y^2} \right)(\sigma_1 + \sigma_2) = 0 \qquad (1.50)$$

This is also known as *Laplace's equation.* It is well known that any function satisfying Eq. (1.50) is called a *harmonic function.* If the boundary is relatively simple, for example, a rectangular region, rigorous mathematical solutions can be obtained by the well-known method of separation of variables. In other cases, either numerical or experimental (analogy) methods can be employed to obtain approximate solutions that are accurate enough for engineering purposes, provided that the boundary values are known.

Experimental methods, or *analogy methods,* use the fact that Laplace's equation serves as the governing equation in many other fields of engineering. An example is that of electrostatic fields in domains enclosed by boundaries where the potential is known, shapes of uniformly stretched membranes, and steady-state temperature distributions. Although the physical quantities are quite different, the mathematical governing equations are exactly the same. We can therefore use one quantity that is easily measured by experiments to study the behavior of other quantities. The electrical-analogy method, for example, is an excellent means of determining $\sigma_1 + \sigma_2$ in the interior of any complicated model under plane stress state, since means are readily available for applying and measuring voltage distributions with sufficient accuracy. It was shown that fringe orders determined by the electrical-analogy method can be as accurate as those obtained by the shear-difference method or by the four-point influence method, which will be discussed shortly. The work involved in the electrical-analogy method is about the same as for the four-point influence method, so the former is not detailed here.

The four-point influence method is one of the numerical methods that can be used very efficiently to solve Laplace's equations. The method makes use of the fact that the value of the function at any point depends directly on the values of the function around its neighborhood and has the added advantage that isochromatic data alone are sufficient for determining $\sigma_1 + \sigma_2$. The domain to be analyzed is covered by a relaxation grid, with the grid lines usually known as *strings* and the grid points referred to as *nodes*. For simplicity, let $\sigma_1 + \sigma_2$ be I (the stress invariant in plane stress cases). The values of I at nodes located on the boundary are given by the boundary conditions. The values of I at nodes located in the domain, called simply the *interior nodes*, can be expressed by the values of I at its four neighborhood nodes, shown in Fig. 4.24, as follows:

$$I_0 = A_1 I_1 + A_2 I_2 + A_3 I_3 + A_4 I_4 \tag{4.68}$$

where constants $A_1, A_2, A_3,$ and A_4 are

$$A_1 = \frac{y_2 y_4 x_3}{(y_2 y_4 + x_1 x_3)(x_1 + x_3)}$$

$$A_2 = \frac{x_1 x_3 y_4}{(y_2 y_4 + x_1 x_3)(y_2 + y_4)}$$

$$\tag{4.69}$$

$$A_3 = \frac{x_1 y_2 y_4}{(y_2 y_4 + x_1 x_3)(x_1 + x_3)}$$

$$A_4 = \frac{x_1 x_3 y_2}{(y_2 y_4 + x_1 x_3)(y_2 + y_4)}$$

When $x_1 = x_3 = y_2 = y_4$, $A_1 = A_2 = A_3 = A_4 = 0.25$, the computation is simplified.

Once the value of I at every interior point has been established using Eqs. (4.68) and (4.69), a set of linear algebraic equations are formed which can be solved by any available method. The method to be illustrated by an example utilizes an iteration procedure by which estimated values of I at interior nodes are systematically improved by employing Eqs. (4.68) and (4.69). The advantage of this method is that any mistake

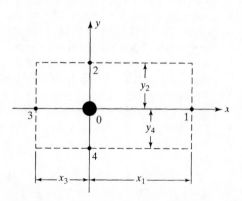

FIGURE 4.24 Grid nodes for a general case.

made during the iteration will not affect the final results. Also, the method is very simple to perform because of the recalculation feature in Lotus-123 [14].

EXAMPLE 4.10

Consider the rectangular model, 8 in. by 6 in., shown in Fig. E4.10. The values of $I = \sigma_1 + \sigma_2$ are known from the isochromatic fringe pattern around the boundary. For simplicity, a square grid is used. The values of I on the boundary nodes in fringe order are also shown in Fig. E4.10. Determine the values of I at all interior nodes using the four-point influence method.

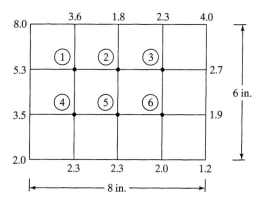

FIGURE E4.10 Grid pattern of a rectangular model.

Solution Let the initial trial values be $I_2^0 = I_3^0 = I_4^0 = I_5^0 = I_6^0 = 2.0$. The initial values are not critical, since they only affect the number of iterations. Then we proceed to improve the values at interior nodes by using Eq. (4.68) until they approach the correct values to within the limits of accuracy set by the accuracy of measurement of the fringe orders at the boundary. Here the values are accurate to one decimal place.

Let a superscript represent the number of iterations and a subscript the number of nodes, as shown in Fig. E4.10. Therefore, we have

$$I_1^1 = \frac{I_2^0 + 3.6 + 5.3 + I_4^0}{4} = \frac{2.0 + 3.6 + 5.3 + 2.0}{4} \approx 3.2$$

$$I_2^1 = \frac{I_3^0 + 1.8 + I_1^1 + I_5^0}{4} = \frac{2.0 + 1.8 + 3.2 + 2.0}{4} \approx 2.3$$

$$I_3^1 = \frac{2.7 + 2.3 + I_2^1 + I_6^0}{4} = \frac{2.7 + 2.3 + 2.3 + 2.0}{4} \approx 2.3$$

$$I_4^1 = \frac{I_5^0 + I_1^1 + 3.5 + 2.3}{4} = \frac{2.0 + 3.2 + 3.5 + 2.3}{4} \approx 2.8$$

$$I_5^1 = \frac{I_6^0 + I_2^1 + I_4^1 + 2.3}{4} = \frac{2.0 + 2.3 + 2.8 + 2.3}{4} \approx 2.4$$

$$I_6^1 = \frac{1.9 + I_3^1 + I_5^1 + 2.0}{4} = \frac{1.9 + 2.3 + 2.4 + 2.0}{4} \approx 2.2$$

Note that the superscript represents the iteration number. We may repeat the procedure above successively using the new values in Eq. (4.68):

$$I_1^2 = \frac{I_2^1 + 3.6 + 5.3 + I_4^1}{4} = \frac{2.3 + 3.6 + 5.3 + 2.8}{4} \approx 3.5$$

$$I_2^2 = \frac{I_3^1 + 1.8 + I_1^2 + I_5^1}{4} = \frac{2.3 + 1.8 + 3.5 + 2.4}{4} \approx 2.5$$

$$I_3^2 = \frac{2.7 + 2.3 + I_2^2 + I_6^1}{4} = \frac{2.7 + 2.3 + 2.5 + 2.2}{4} \approx 2.4$$

$$I_4^2 = \frac{I_5^1 + I_1^2 + 3.5 + 2.3}{4} = \frac{2.4 + 3.5 + 3.5 + 2.3}{4} \approx 2.9$$

$$I_5^2 = \frac{I_6^1 + I_2^2 + I_4^2 + 2.3}{4} = \frac{2.2 + 2.5 + 2.9 + 2.3}{4} \approx 2.5$$

$$I_6^2 = \frac{1.9 + I_3^2 + I_5^2 + 2.0}{4} = \frac{1.9 + 2.4 + 2.5 + 2.0}{4} \approx 2.2$$

Similarly,

$$I_1^3 = \frac{I_2^2 + 3.6 + 5.3 + I_4^2}{4} = \frac{2.5 + 3.6 + 5.3 + 2.9}{4} \approx 3.6$$

$$I_2^3 = \frac{I_3^2 + 1.8 + I_1^3 + I_5^2}{4} = \frac{2.4 + 1.8 + 3.6 + 2.5}{4} \approx 2.6$$

$$I_3^3 = \frac{2.7 + 2.3 + I_2^3 + I_6^2}{4} = \frac{2.7 + 2.3 + 2.6 + 2.2}{4} \approx 2.5$$

$$I_4^3 = \frac{I_5^2 + I_1^3 + 3.5 + 2.3}{4} = \frac{2.5 + 3.6 + 3.5 + 2.3}{4} \approx 3.0$$

$$I_5^3 = \frac{I_6^2 + I_2^3 + I_4^3 + 2.3}{4} = \frac{2.2 + 2.6 + 3.0 + 2.3}{4} \approx 2.5$$

$$I_6^3 = \frac{1.9 + I_3^3 + I_5^3 + 2.0}{4} = \frac{1.9 + 2.5 + 2.5 + 2.0}{4} \approx 2.2$$

If the iteration above is repeated one more time, we will find that further corrections do not alter the values within the limit of error of the given boundary values. Thus the final results at the six interior nodes are, respectively, 3.6, 2.6, 2.5, 3.0, 2.5, and 2.2. The principal stresses σ_1 and σ_2 also can be obtained at the six interior nodes.

If Quatro or Lotus-123 is available, the example above can be finished within 5 minutes by using the recalculation function. The sample output is shown below.

Four-Point Influence Method			
8	3.6	1.8	2.3	4
5.3	3.622153	2.601242	2.455486	2.7
3.5	2.987370	2.527329	2.220703	1.9
2	2.3	2.3	2	1.2

NOTE: Numbers in dotted box are interior points.

4.11   RELATIONSHIP BETWEEN STRESSES IN MODELS AND ACTUAL STRUCTURES

So far, we have concentrated on determining stresses by using the photoelastic method in the model of a structure. How do these stresses relate to the stresses in the actual structure? Fortunately, it can been shown that for many practical problems in two dimensions, stresses depend only on the geometry and applied forces, not on the physical properties of the material. Recall that the compatibility equation can be expressed in terms of stresses in a plane stress case as follows:

$$\left(\frac{\partial^2}{\partial x^2} + \frac{\partial^2}{\partial y^2} \right)(\sigma_x + \sigma_y) = -(1 + \mu)\left(\frac{\partial X}{\partial x} + \frac{\partial Y}{\partial y} \right) \qquad (1.51)$$

It is clear that Eq. (1.51) is independent of Young's modulus E and also independent of Poisson's ratio (with a few exceptions, such as multiconnected domains), provided that the body forces are either absent or constant (e.g., gravitational fields). Thus the stress distribution obtained for a plane stress case by a photoelastic analysis can be applied to a prototype made from any material.

Therefore, it is only necessary that a model be geometrically similar to the prototype and that the applied loads be distributed the same way as in the actual structure. The stresses in the model can then be used to determine the stresses in the actual structure by the process of multiplying by a scaling factor obtained from the laws of similitude. For example, for a two-dimensional case, the stress σ in a structure can be expressed by

$$\sigma = \frac{F L_m^2}{F_m L^2} \sigma_m \qquad (4.70)$$

where F is a force applied to the structure, L is a linear dimension in the structure, and m = a subscript denoting the corresponding quantities in the photoelastic model.

In practice, two-dimensional models, which are geometrically similar to the structural part except for thickness, are often used. In such cases, the stress σ in the prototype can be expressed by

$$\sigma = \frac{F L_m d_m}{F_m L d} \sigma_m \qquad (4.71)$$

where d is the thickness of the structural part. In some cases, a mismatch of Poisson's ratio between the photoelastic model material and the structural material does affect the results in real structures. Fortunately, most photoelastic materials have a Poisson's ratio in the range 0.3 to 0.35, which is close to that of structural metals. Also, the mismatch of Poisson's ratio has little influence on the maximum principal stress, which is of most interest in engineering stress analysis.

4.12   MATERIALS FOR PHOTOELASTIC MODELS

An ideal photoelastic model material should meet certain requirements, which are summarized as follows [1,2,4,7]:

1. The material should be isotropic (both mechanical and optical) and homogeneous except where anisotropic properties are required; it should be transparent to visible light and relatively colorless.
2. The material should have a linear relationship between stress and strain, and between stress and optical response, and it should be relatively free from creep effects within the working time period.
3. The material should have a minimum time-edge effect, which results primarily from absorption of moisture at the machined edges. The induced stress on the edges produces fringes when no loads are applied.
4. The material should be easy to machine without inducing stress or chipping. Usually, high cutting forces should be avoided in machining photoelastic models.
5. The material should be free of any residual stress. In certain cases, the residual stress can be reduced to a minimum by careful annealing of the material.
6. The material should be sensitive enough to provide comparatively more fringe numbers under relatively low loads. In other words, the material fringe value, M_f^σ, should be low.
7. Young's modulus should be as high as possible so that the shape of the model, when deformed under applied loads, will not differ appreciably from that of the structural element. Keep in mind that a low Young's modulus for the model will introduce large error in the experimental results.
8. The proportional limit, σ_{pl}, should also be as high as possible to produce a higher-order fringe pattern to improve the accuracy of stress determination, since the model can be loaded to a higher level without endangering the safety of the model.

Among these requirements, items 6 to 8 are the most important ones. Quantities called the *figure of merit* and the *sensitivity index* are introduced and defined by [1, 2]

$$Q = \frac{E}{M_f^\sigma} \qquad \text{in.}^{-1} \text{ (or mm}^{-1})$$

$$S = \frac{\sigma_{pl}}{M_f^\sigma} \qquad \text{in.}^{-1} \text{ (or mm}^{-1})$$

(4.72)

Both Q and S should be as high as possible. In practice, no ideal model material exists to perfectly satisfy all the requirements mentioned above. Typical model materials are epoxy resins, CR-39 (or Homolite 911), polycarbonate, Homolite 100, plexiglass (or perspex), glass, and urethane rubber. Their properties at room temperature are given in Table 4.3.

TABLE 4.3   Approximate Properties of Photoelastic Model Materials at Room Temperature

Materials	Stress Fringe Value M_f (λ = 18.5µin. or 470 nm) [lb/in. (kN/m)]	Young's Modulus E ksi (MPa)	Poisson's Ratio μ	Proportional Limit, σ_{pl} [ksi (MPa)]	Sensitivity Index $S = \sigma_{pl}/f_\sigma$ in.$^{-1}$ (mm)	Figure of Merit $Q = E/f_\sigma$ in.$^{-1}$ (mm)	Time-Edge Effects	Creep	Machinability	Remarks
Epoxy resins	52 (9.1)	475 (3,275)	0.37	8 (55.2)	154 (6.1)	9,100 (360)	Good	Good	Good	Widely used for both two- and three-dimensional analysis
CR-39 (Homolite 911)	77 (13.5)	250 (1,725)	0.42	3 (20.7)	39 (1.53)	3,250 (128)	Poor	Poor	Poor	Normal-incident analysis only
Polycarbonate	34 (6.0)	360 (2,480)	0.38	5 (34.5)	147 (5.75)	10,600 (410)	Excellent	Excellent	Poor	Difficult to machine, expensive
Homolite 100	120 (21)	560 (3,860)	0.35	7 (48.3)	58 (2.3)	4,700 (184)	Excellent	Excellent	Good	Extremely brittle, edge chipping
Urethane rubber	0.9 (1)	0.45 (3.1)	0.46	0.02 (0.14)	22 (0.88)	500 (19)	Excellent	Excellent	Poor	Used for demonstration; dynamic modeling problems by self-weight
Plexiglass (U.S.) or Perspex (U.K.)	600 (105)	400 (2,760)	0.38	Not available	Not available	670 (26)	Excellent	Excellent	Good	Used more extensively with the development of new technology
Glass	1,350 (241)	9,000 (62,000)	0.25	8.7 (60.0)	6.3 (0.25)	6,500 (260)	Excellent	Excellent	Very poor	Extremely brittle, chipping at loading points

Source: References 1, 2, 4, and 7.

A variety of epoxy resins are generally used as model material for both two- and three-dimensional photoelastic applications. They have good mechanical and optical properties and are easy to machine, and their time effects may be kept fairly small by controlling the humidity conditions. CR-39 has good optical sensitivity and is highly transparent and colorless, but it is rather brittle. Due to high initial stress in the thickness direction, produced during the manufacturing process, it is suitable only for normal incidence analysis and general two-dimensional analysis.

Polycarbonate has both a high figure of merit (10,600 in.$^{-1}$ for blue light) and a high sensitivity index (147 in.$^{-1}$ for blue light). It exhibits little creep at room temperature and is fairly free of time-edge effects. Unfortunately, it is relatively difficult to machine. Like polycarbonate, Homolite 100 exhibits little creep at room temperature and is fairly free of time-edge effects. It can be machined by routing although it is very brittle. Due to a low figure of merit (4700 in.$^{-1}$ for blue light) and sensitivity index (58 in.$^{-1}$ for blue light), higher fringe orders cannot be obtained.

Urethane rubber is popular for dynamic stress analysis due to its low Young's modulus value (450 psi), which results in low-velocity stress waves, thus simplifying the problem of recording fringe patterns. It is used for demonstration models and for modeling problems loaded by self-weight.

Plexiglass is highly transparent and colorless. Due to its high stress fringe value (600 lb/in. for blue light), it is sometimes used in composite models with a more optically sensitive material, so that the polarized light passes through parts made of plexiglass without being affected by the existing stresses in them. The material will probably be used more extensively with the development of new techniques, such as computer-aided photoelasticity. Glass was the first model material used in photoelasticity. Although it has many advantages, its use is limited due to the fact that it is very difficult to shape. In recent years it has undergone a resurgence of popularity as a photoelastic material for some special applications, such as in long-life optical transducers in civil engineering structures.

4.13   ADVANCED PHOTOELASTICITY

In practice, many problems are three dimensional in nature. Thus techniques in three-dimensional photoelasticity are rapidly developing. Among the most powerful of these is the frozen-stress method. The extraordinary feature of three-dimensional photoelasticity is that it provides a unique opportunity to determine the stress distributions inside a body. Certain photoelastic model materials, such as epoxy resins, exhibit the *frozen-stress phenomenon*. This phenomenon occurs when the material is heated to its stress annealing (or stress freezing) temperature, stressed (usually by dead-weight loads) after the temperature has reached its equilibrium in the material, and then cooled slowly to room temperature with the loads still applied. The optical response, related to the mechanical stress, remains fixed in the material after removal of the loads at room temperature.

Furthermore, the optical response is not disturbed even when the material is cut into thin slices. This phenomenon forms the basis of the frozen-stress method, widely employed in three-dimensional photoelasticity. The slices can be analyzed by the polariscope, employing essentially the same technique as that developed for two-dimensional photoelasticity. Constitutive equations in three dimensions can be used to determine the stress state at any point on the surface directly from optical measurements. However, for determination of stress at an interior point, a supplementary mathematical procedure is necessary, just as for two-dimensional cases.

The techniques described for two-dimensional photoelasticity are equally applicable to dynamic loadings. Perhaps the most difficult problem encountered in dynamic photoelasticity is the recording of fringe patterns, since the propagation velocity of fringes is extremely high. The Cranz–Schardin spark camera is one high-speed recording system used in studying wave propagation and dynamic fracture. New techniques of photoelasticity have been developed, such as computer-aided photoelasticity, methods of integrated photoelasticity, and holo-photoelastic methods. Although the traditional areas of photoelastic applications have been partially replaced by numerical methods with the development of computers, photoelastic applications have also expanded. For example, photoelastic methods have recently been used in experimental fracture mechanics, and photoelastic methods remain an important stress analysis tool.

The topics of three-dimensional photoelasticity, dynamic photoelasticity, and new techniques and applications are too complex to be described in this book. The interested reader is directed to references [1, 2, 3, 12, & 13] for further reading.

PROBLEMS

4.1. The frequency of a monochromatic light is 7.1×10^{14} Hz. Determine **(a)** the wavelength of this light in air, **(b)** the wavelength of this light in a glass plate ($i_b \approx 1.522$), and **(c)** the velocity of propagation in the glass plate.

4.2. A monochromatic incident ray strikes a glass plate at an angle of incidence of α. The ray emerges from the other side of the plate parallel to its initial direction but with a transverse displacement, as shown in Fig. P4.2. The thickness of plate is d and $i_b = 1.5$. **(a)** Derive an expression for h in terms of α, d, and i_b. **(b)** If $\alpha = 30°$ and $d = 30$ mm, determine h.

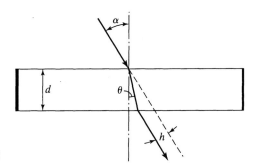

FIGURE P4.2

4.3. Determine the linear phase shift δ in Problem 4.2 after the light has passed through the glass plate.

4.4. Prove that Eq.(4.16) is valid.

4.5. A half-wave plate is designed for operation at λ = 548.0 nm. Determine the relative angular phase shift if it is operating with a light (λ = 590.0 nm).

4.6. If the fast axes of the quarter-wave plates in Example 4.4 are parallel to each other, what conclusion can you make based on similar derivations?

4.7. At a particular point in a photoelastic model, a fringe order of 4.5 is established in a polariscope with a Mercury green light (λ = 546.1 nm). What fringe order will be observed if the light source is changed to a helium–neon red light (λ = 632.8 nm)?

4.8. In Problem 4.7, the material fringe value is 65 lb/in., which is established with a Mercury green light (λ = 546.1 nm). What is the material fringe value if a helium–neon red light (λ = 632.8 nm) is employed?

4.9. If the axes for the polarizer and analyzer are parallel in a plane polariscope, derive an expression for the light components that are passed through the stressed model and transmitted by the analyzer. Also determine the conditions for zero intensity of the light ($I = 0$).

4.10. Derive equations for light passing through a stressed model in a circular polariscope (the axes of polarizer and analyzer are crossed but the axes of the two quarter-wave plates are parallel). Discuss the conditions when extinction occurs ($I = 0$).

4.11. Derive equations for light passing through a stressed model in a circular polariscope (the axes of polarizer and analyzer are parallel and the axes of the two quarter-wave plates are also parallel). Discuss the conditions when extinction occurs ($I = 0$).

4.12. Derive equations for light passing through a stressed model in a circular polariscope (arrangement 1) if the quarter-wave plates have some imperfection. Assume that $\Delta = \pi/2 - \epsilon$ for both plates. Discuss the conditions when extinction occurs ($I = 0$) and the effect due to the imperfection ϵ.

4.13. Repeat Problem 4.12 if arrangement 4 is employed.

4.14. Verify Eq. (4.43).

4.15. The following procedures have been employed to establish the fractional fringe order at a point in a stressed model.

   (a) Begin with a dark-field plane polariscope with white light, then rotate the crossed polarizer and analyzer together until the black isoclinic fringe covers the point.

   (b) Insert two quarter-wave plates to form a dark-field circular polariscope (arrangement 1) with monochromatic light.

   (c) Rotate the polarizer alone until a fringe (order of n) in the neighborhood is coincident with the point.

   Verify that the effect is the same as in the Tardy method.

4.16. The following method is an alternative way to establish the fractional fringe order at a point in a stressed model.

   (a) Begin with a dark-field plane polariscope and position the stressed model so that the principal direction of stresses at a point of interest is oriented ±45° with respect to the polarizer axis.

   (b) Insert a wave plate between the stressed model and analyzer with its fast axis parallel to the polarizer axis.

   (c) Rotate the analyzer so that extinction occurs. Verify the validity of this method using equations.

4.17. Figure P4.17 shows a photograph of isochromatic patterns for two beams in a dark-field circular polariscope. The heights for the upper beam and lower beam are 0.471 and 0.467 in.; the beam thicknesses are respectively, 0.241 and 0.239 in. C is approximately

0.3 in. The photograph is taken at load $P = 20$ lb. Determine the material fringe value M_f for both beams.

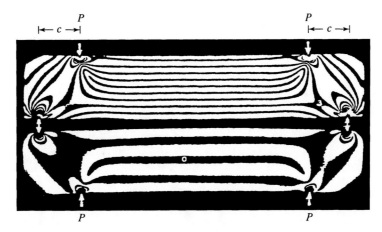

FIGURE P4.17 [5]

4.18. Figure P4.18 shows the isochromatic fringe patterns (both dark-field and light-field) through a section of square tube with a circular bore subjected to an internal pressure. Plot the fringe orders (both integer orders and half orders) versus the position along the x axis from $x = r$ to $x = a/2$ in Fig. P4.18. ($r = 38.5$ mm, $a = 90$ mm).

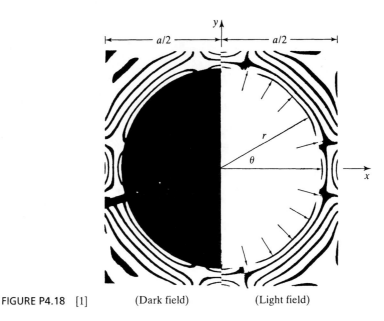

FIGURE P4.18 [1]   (Dark field)   (Light field)

4.19. Plot the fringe orders (both integer orders and half orders) versus the position along the outside boundary [i.e., from point ($x = a/2, y = 0$) to point ($x = a/2, y = a/2$) in Fig. P4.18]. Assume that $b = 90$ mm.

4.20. In Fig. P4.18, the internal pressure $P = 200$ psi, material fringe value $M_f^\sigma = 60$ lb/in., and thickness is 5.08 mm. Plot the stress distribution σ_1 along the inner boundary from $\theta = 0$ to $\theta = \pi/2$.

4.21. Figure P4.21 is the isochromatic pattern of a centrally loaded arch (simply supported). Determine whether the photo is a dark- or light-field isochromatic pattern. State your reason clearly. Next determine the fringe orders at points A, B, C, and D.

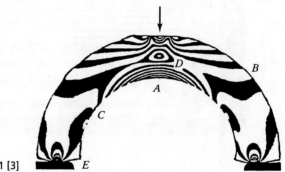

FIGURE P4.21 [3]

4.22. Figure P4.22 shows the isochromatic pattern of shallow grooves in pure bending. Plot the fringe orders versus the position along section A–A. Assume that the height at A–A is 32 mm.

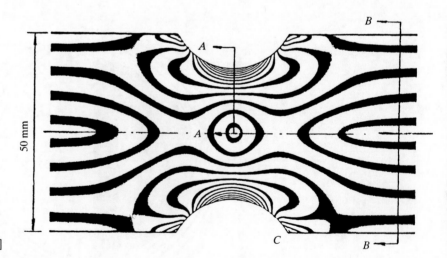

FIGURE P4.22   [5,13]

4.23. Determine the stress concentration factor in Problem 4.22. The height of the beam is 50 mm, the height at section A–A is 32 mm, and the thickness is d.

4.24. Figure P4.24 is the isochromatic pattern for a beam under three-point loading. The fringe orders along the upper edge are also shown in the figure. Plot the fringe orders versus the distance along the lower edge. The beam length between two supports is assumed to be 76.2 mm.

FIGURE P4.24   [5,13]

4.25. In Eq.(4.56), N_α is the fringe order obtained from the oblique-inclined method by rotating the model about the σ_1 axis. Derive the corresponding equation if N_α is determined by rotating the model about the σ_2 axis.

4.26. Figure P4.26 shows the isoclinic and isochromatic fringes along the x axis for a square plate subjected to a pair of concentrated loads along the diagonal line. Calculate the stresses (σ_x, σ_y, and τ_{xy}) at points 0, 1, 2, 3, 4, 5, 6 on the x axis by the shear-difference method and plot the stress σ_x, σ_y, and τ_{xy} distribution along line KO (the length is 14.25 mm). For simplicity, assume that $|\Delta x/\Delta y| = 1$ (Δx and Δy are constants). All necessary data, given in the tables below, were determined from the isoclinic and isochromatic patterns shown in Fig. P4.26.

(a) Along line AB

Point:	0	1	2	3	4	5	6
N	1.2	1.44	2	3	3.9	4.6	4.95
θ	45°	33°	27°	20°	13.5°	7.5°	0°

(b) Along line OK

Point:	0	1	2	3	4	5	6
N	0.80	1.29	2.03	2.89	3.73	4.47	4.80
θ	45°	28°	23.4°	18.2°	12°	6.3°	0°

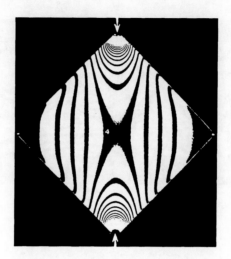

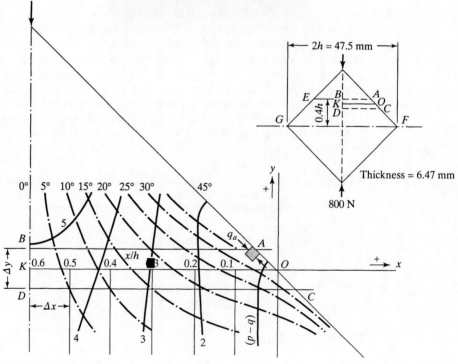

FIGURE P4.26 [13]

(c) Along line *CD*

Point:	0	1	2	3	4	5	6
N	0.66	1.4	2.2	2.8	3.7	4.3	4.7
θ	45°	23°	18.7°	14.5°	10°	5°	0°

Note that θ is the angle between σ_1 and *X*, measured in a counterclockwise direction.

4.27. Point *P*, shown in Fig. P4.27, is on the stress-free boundary. To determine whether the remaining nonzero principal stress is σ_1 or σ_2, a small normal pressure *p* (*p* > 0) is applied at point *P*. It is found that the fringe order is increased from 3 to 4. Prove that the nonzero stress is σ_1.

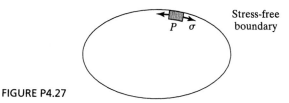

Stress-free boundary

FIGURE P4.27

4.28. Consider the rectangular model, 12 in. by 8 in, shown in Fig. P4.28. The boundary values of the stress invariant $(\sigma_1 + \sigma_2)$ are also shown in the figure, which is in fringe order. Determine the values of $\sigma_1 + \sigma_2$ at all interior nodes by using the four-point influence method. (*Hint:* Use Lotus-123.)

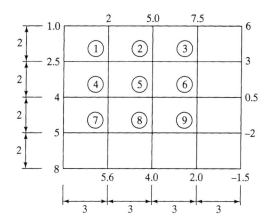

FIGURE P4.28

REFERENCES

[1] J. W. Dally and W. F. Riley, *Experimental Stress Analysis,* 2nd ed., McGraw-Hill, New York, 1978.

[2] C. P. Burger, Photoelasticity, in A. S. Kobayashi, ed., *Handbook on Experimental Mechanics,* Second Edition, VCH Publishers, Inc., New York, 1993.

[3] A. S. Kobayashi, ed., *Manual of Engineering Stress Analysis,* 3rd ed., Prentice Hall, 1982.

[4] G. S. Holister, *Experimental Stress Analysis: Principles and Methods,* Cambridge University Press, New York, 1967.

[5] M. M. Leven, ed., *Photoelasticity: The Selected Scientific Papers of M. M. Frocht,* Pergamon Press, Elmsford, N.Y., 1969.

[6] A. Kuske and G. Robertson, *Photoelastic Stress Analysis,* Wiley, New York, 1974.

[7] R. B. Heywood, *Photoelasticity for Designers,* Pergamon Press, Elmsford, N.Y., 1969.

[8] J. C. Maxwell, On the equilibrium of elastic solids, *Trans. R. Soc. Edinburgh,* **XX,** Pt. 1, 1853.

[9] J. W. Dally and F. J., Ahimaz, Photographic method to sharpen and double isochromatic fringes, *Exp. Mech., ***2,** 170, 1962.

[10] M. H. L. Tardy, Méthode pratique d'examen de desure de la biréfringence des verres d'optique, *Rev. Opt., ***8,** 59, 1929.

[11] *Introduction to Stress Analysis by the PhotoStress Method,* Tech Note, TN-702-1, Measurements Group, Inc., Raleigh (North Carolina), 1989.

[12] M. M. Frocht, *Photoelasticity,* Vol. I, Wiley, New York, 1948 (3rd printing, 1961).

[13] M. M. Frocht, *Photoelasticity,* Vol. II, Wiley, New York, 1948 (3rd printing, 1961).

[14] *1-2-3 Access System* (software), Release 2, Lotus Development Corporation, 1985.

[15] D. C. Drucker, The method of oblique incidence of photoelasticity, *Proc. SESA, ***8,** 51, 1950.

[16] G. L. Cloud, *Optical Methods of Engineering Analysis,* Cambridge University Press, 1995.

C H A P T E R   5

Photoelastic-Coating Method

5.1 INTRODUCTION

The photoelastic-coating method extends photoelasticity to direct measurement of surface strains on a structural component. A thin sheet of photoelastic material is bonded directly with a reflective cement to the well-polished surface of the structural component being analyzed. When the structural component is loaded, the photoelastic coating will be deformed, along with the surface of the structural component, so that a strain field is developed in the coating. By employing a reflection polariscope, the fringe orders can be recorded and then used to determine the difference in the principal strains (or stresses) on the surface of the structural component.

The photoelastic-coating method has many advantages over other methods in experimental stress analysis. Like the photoelasticity method, it provides a full-field strain (stress) distribution on the surface of the structure. This is very useful, for example, in determining the position and value of maximum stress and stress concentrations on a free surface of a three-dimensional body. Like the strain gage measurement method, it is nondestructive and can directly measure the strains on the surface of a structural component. The photoelastic-coating method is also very helpful in analyzing problems when the stress–strain relationships of the structural materials exhibit complex nonlinear, elastoplastic, or anisotropic behaviors. This can be done by proper selection of coating materials to convert the above-mentioned problems in the test object into a relatively simple linear elastic problem in the isotropic coating. Reference [10] illustrates some examples of industrial and research applications by using photoelastic coatings. It should be mentioned that the basic principles and techniques described in Chapter 4 are equally valid for photoelastic coatings.

5.2   REFLECTION POLARISCOPE

Figure 5.1 shows two basic arrangements of the reflection polariscopes that are commonly used in photoelastic-coating analysis by the normal incidence method. Figure 5.1a shows the most popular type of simple portable polariscope. The relative positions of polarizer, quarter-wave plates, and analyzer are similar to the circular polariscope described in Chapter 4. As shown in Fig. 5.1a, the incident light is not exactly normal to the coating surface, so that the reflected light in the coating is not in the same line as the incident light. In general, the error introduced due to this slight obliquity of the incident light is negligible, provided that the polariscope is kept more that 2 ft away from the coating. However, when the strain gradient is high, a large error may be introduced in the recorded data by employing this type of reflection polariscope.

The polariscope shown in Fig. 5.1b can be employed when the strain gradient on the surface is high. Since the incident and reflected lights are perpendicular to the surface of the coating, the results obtained are more accurate. However, due to the reflections by the partial mirror, the light intensity is relatively weaker, and nonlinearity may be produced during compensation.

Figure 5.2 illustrates some specialized photoelastic equipment [1,11] that can be used in the photoelastic-coating method. Figure 5.2a shows the basic reflection polariscope (Model 031), which consists of three parts: the special light source (part 3), the assembly to receive the light (part 1), and the assembly to provide the measurement scale (part 2). This polariscope is also equipped to accept various accessories: for example, Model 031 equipped with a camera (Fig. 5.2b) and Model 030 equipped with a telemicroscope (Fig. 5.2c). Figure 5.2d illustrates the direct digital display reflection polariscope.

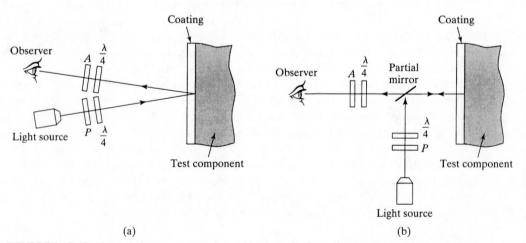

(a)   (b)

FIGURE 5.1   Reflection polariscopes commonly used in photoelastic-coating measurements: P, polarizer; A, analyzer; $\lambda/4$, quarter-wave plate.

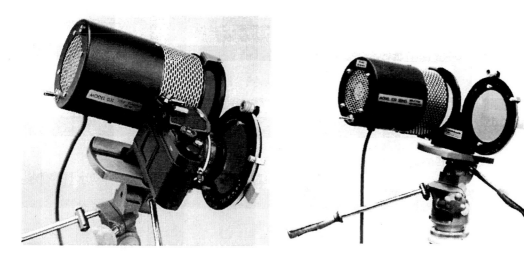

FIGURE 5.2   Specialized photoelastic equipment (Measurements Group, Inc.).

5.3   BASIC PRINCIPLES OF THE PHOTOELASTIC-COATING APPLICATIONS

Consider a specimen with a thin layer of coating, as shown in Fig. 5.3. Both the specimen and the coating are assumed to be in a plane stress state, since a thin photoelastic coating is usually bonded to the specimen surface when it is in an unstressed stress state. Let the z axis be in the thickness direction and the xy plane in the specimen surface (or coating) plane. Thus the nonzero stress components in the specimen and the coating are $\sigma_x^s, \sigma_y^s, \tau_{xy}^s, \sigma_x^c, \sigma_y^c$, and τ_{xy}^c; the nonzero strain components are $\epsilon_x^s, \epsilon_y^s, \epsilon_z^s, \gamma_{xy}^s, \epsilon_x^c$, $\epsilon_y^c, \epsilon_z^c$, and γ_{xy}^c, where the superscripts s and c denote the specimen surface (or simply specimen) and the coating, respectively.

Assume that the bond is perfect so that there is no loss or amplification during transmission of the surface displacement of the specimen to the coating via the adhesive. At the interface, we have

$$\epsilon_x^c = \epsilon_x^s \qquad \epsilon_y^c = \epsilon_y^s \qquad \gamma_{xy}^c = \gamma_{xy}^s \tag{5.1a}$$

or

$$\epsilon_1^c = \epsilon_1^s \quad \text{and} \quad \epsilon_2^c = \epsilon_2^s \tag{5.1b}$$

where subscripts 1 and 2 represent the principal directions. Equations (5.1) represent the strain relationships between the specimen and the coating. Furthermore, the strains in the coating are assumed constant through the thickness, since the coating is relatively thin.

Assume that the specimen material is isotropic and linear elastic. By using Eqs. (1.46) and (5.1b), we have

$$\epsilon_1^s = \frac{1}{E^s}(\sigma_1^s - \mu^s \sigma_2^s) = \epsilon_1^c = \frac{1}{E^c}(\sigma_1^c - \mu^c \sigma_2^c)$$
$$\epsilon_1^s = \frac{1}{E^s}(\sigma_2^s - \mu^s \sigma_1^s) = \epsilon_2^c = \frac{1}{E^c}(\sigma_2^c - \mu^c \sigma_1^c) \tag{5.2}$$

where E and μ are Young's modulus and Poisson's ratio, respectively. Equation (5.2) can be used to get

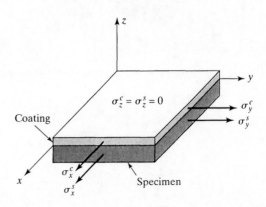

FIGURE 5.3   Specimen with a thin layer of coating.

$$\sigma_1^s = \frac{E^s}{E^c\,(1 - \mu^{s2})}\left[(1 - \mu^c\mu^s)\sigma_1^c - (\mu^c - \mu^s)\sigma_2^c\right]$$

$$\sigma_2^s = \frac{E^s}{E^c\,(1 - \mu^{s2})}\left[(1 - \mu^c\mu^s)\sigma_2^c - (\mu^c - \mu^s)\sigma_1^c\right]$$

(5.3)

which are the stress relationships between the specimen and the coating. The stress field in the specimen can be determined by Eqs. (5.3) once the stress field in the coating is known. From Eqs. (5.3) we can obtain the principal stress difference in the coating, given by

$$\sigma_1^c - \sigma_2^c = \frac{E^c\,(1 + \mu^s)}{E^s\,(1 + \mu^c)}\,(\sigma_1^s - \sigma_2^s)$$

(5.4)

It is clear that the stresses (or stress difference) in the coating are linearly related to the stresses (or stress difference) in the specimen.

5.4 STRESS- OR STRAIN-OPTIC LAW FOR COATINGS

In Chapter 4 the stress-optic law for a photoelastic material in plane stress was expressed as

$$\sigma_1 - \sigma_2 = \frac{NM_f^\sigma}{d}$$

(4.18)

where σ_1 and σ_2 are principal stresses, and N, M_f^σ, and d are the isochromatic fringe order, the material stress fringe value or simply the material fringe value, and the specimen thickness, respectively. Equation (4.18) can also be applied to the coating material. Note that the light transmits through the thickness of the coating twice (Fig. 5.1), thus

$$\sigma_1^c - \sigma_2^c = \frac{NM_f^\sigma}{2d^c}$$

(5.5)

However, it is sometimes preferable to use the strain-optic law since the coating material is usually different from the specimen material. From Eq. (5.2) we have

$$\epsilon_1^s - \epsilon_2^s = \epsilon_1^c - \epsilon_2^c = \frac{(1 + \mu^c)(\sigma_1^c - \sigma_2^c)}{E^c}$$

Substituting Eqs. (5.5) into the equation above results in

$$\epsilon_1^s - \epsilon_2^s = \epsilon_1^c - \epsilon_2^c = \frac{1 + \mu^c}{E^c}\frac{NM_f^\sigma}{2d^c}$$

(5.6)

or

$$\epsilon_1^s - \epsilon_2^s = \epsilon_1^c - \epsilon_2^c = \frac{NM_f^\epsilon}{2d^c} = NC_f^\epsilon$$

where $M_f^\epsilon = (1 + \mu^c)M_f^\sigma/E^c$ is the material strain fringe value and $C_f^\epsilon = M_f^\epsilon/2d^c$ is the coating fringe value (or coating sensitivity), taking into account the thickness of the coating. Equation (5.6) is called the *strain-optic law*.

If the thickness of coating is not relatively thin, or if the modulus of the specimen material is relatively low, a correction must be made in the strain-optic law. This is necessary because the coating itself may carry a noticeable portion of the loading (known as the *reinforcing effect*), or the nonuniform strain distribution through the coating thickness may be significant, such as in the case of bending loading. Taking these effects into account and noting that the optical response of the coating is related to the average strain, Eq. (5.6) becomes

$$(\epsilon_1^s - \epsilon_2^s)_{\text{true}} = C_i (\epsilon_1^c - \epsilon_2^c)_{\text{av}} = (C_i N)C_f^\epsilon \qquad (5.7)$$

where C_i is the correction factor; the subscript "true" denotes the true strains at the surface of the specimen without coatings, and the subscript "av" represents the average strains in the coating.

It can be shown [2] that when the coated specimen or structural component is subjected to a biaxial stress loading (plane stress case), the corresponding correction factor, denoted by C_p, is given by

$$C_p = 1 + \frac{d^c E^c(1 + \mu^s)}{d^s E^s(1 + \mu^c)} \qquad (5.8a)$$

where d, E, and μ are thickness, Young's modulus, and Poisson's ratio for the coating (with the superscript c) or for the specimen (with the superscript s). A plot of Eq. (5.8.a) is given in Fig. 5.4 for four specimens: steel ($E^s = 30,000$ ksi or 207 GPa), aluminum ($E^s = 10,000$ ksi or 69 GPa), magnesium ($E^s = 6,500$ ksi or 45 GPa), and hard vinyl ($E^s = 0.42$ ksi or 2.9 GPa), coated with a typical photoelastic material (e.g., PL-1). The Young's modulus and Poisson's ratio for the coating are 420 ksi (or 2.9 GPa) and 0.36. It can be seen from Figure 5.4 that C_p is always greater than unity because the reinforcing effects of the coating always reduce the observed fringe orders.

If the coated plate is subjected to bending, in addition to the reinforcing effects, the neutral plane will be slightly shifted due to the presence of the coating, and the coating stress in the thickness direction is no longer a constant. It can also be shown [2] that the correction factor for plates or beams in bending, denoted by C_b, is given by

$$C_b = \frac{1 + \alpha\beta}{1 + \beta} \left[4(1 + \alpha\beta^3) - \frac{3(1 - \alpha\beta)^2}{1 + \alpha\beta} \right] \qquad (5.8b)$$

where

$$\alpha = \frac{(1 - \mu^{s2})E^c}{(1 - \mu^{c2})E^s} \qquad \beta = \frac{d^c}{d^s} \qquad (5.8c)$$

Equation (5.8.b) is shown graphically in Fig. 5.5 for the same specimen materials (steel, hard vinyl, aluminum, and magnesium) coated with the photoelastic material PL-1. It can be seen from Fig. 5.5 that C_b is less than unity for high Young's modulus materials,

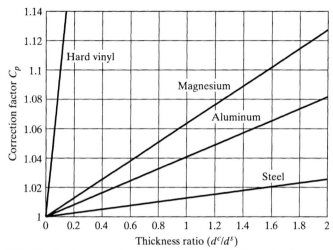

FIGURE 5.4 Correction factor C_p versus ratio of thickness in plane stress.

because the strain-magnification effect, which occurs because of the strain gradient through the thickness of coating, is predominant. The observed fringe order must therefore be multiplied by a factor of less than unity. However, for low Young's modulus specimen materials [e.g., reinforced plastic (E^s = 3 ksi or 21 GPa) and vinyl (E^s = 0.42 ksi or 2.9 GPa)] the correction factor C_b is greater than 1 (as shown in Fig. 5.5) because the stiffening effect of the coating is predominant. In addition to their uses in correcting fringe-order measurements, Figs. 5.4 and 5.5 may also be used as a guide in selecting a coating thickness. Details for coating selections are discussed in Section 5.7.

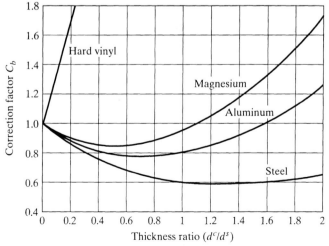

FIGURE 5.5 Correction factor C_b versus ratio of thickness in bending.

5.5 CALIBRATION OF PHOTOELASTIC COATING

For greater accuracy it is necessary to calibrate the coating fringe value C_f^ϵ for each sheet of photoelastic plastic. Theoretically, calibration methods described in Section 4.8 can be employed to determine the coating fringe value C_f^ϵ. However, it is sometimes more convenient to bond a small piece of coating, cut from the same sheet of photoelastic plastic as that to be used on the test object, on a standard cantilever beam subjected to a dead weight loading, as illustrated in Fig. 5.6. By knowing the applied load, the stress and strain fields can be calculated. F_ϵ can therefore be determined by using Eqs. (5.7) or (5.8b) once the fringe order N is known.

EXAMPLE 5.1

Consider an aluminum cantilever beam, which is 0.2 in. thick and 0.75 in. wide and is bonded with a piece of 0.05-in.-thick (0.75-in.-wide) coating on its upper surface, as shown in Fig. 5.6. With a 15-lb applied load, a fringe order of 2 at 7 in. away from the load point is found. Determine the coating fringe value F_ϵ. Young's modulus and Poisson's ratio are 10,000 ksi and 0.33 for the aluminum, and 360 ksi and 0.35 for the coating.

Solution   For an uncoated beam, the stress at the measurement point, located on the upper surface of the beam, is given by

$$\sigma_1^s = \frac{6M}{bd^{s2}} = \frac{(6)(15)(7)}{0.75(0.2)^2} = 21.0 \text{ ksi}$$

$$\sigma_2^s = 0$$

From Eq. (5.2),

$$(\epsilon_1^s - \epsilon_2^s)_{\text{true}} = \frac{(1 + \mu^s)(\sigma_1^s - \sigma_2^s)}{E^s}$$

$$= \frac{(1 + 0.33)(21 - 0)}{10,000} = 2793 \ \mu\text{in./in.}$$

The thickness ratio is $\beta = 0.05/0.2 = 0.25$. Also,

$$\alpha = \frac{(1 - 0.33^2)(360)}{(1 - 0.35^2)(10,000)} = 0.0366$$

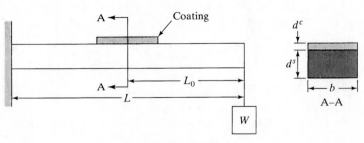

FIGURE 5.6   Cantilever beam calibration method.

Therefore, by Eq. (5.8b), the correction factor for a beam in bending can be calculated as

$$C_b = \frac{1 + (0.0366)(0.25)}{1 + 0.25} \left\{ (4)[1 + (0.0366)(0.25)^2] - \frac{(3)[1 - 0.0366(0.25)^2]^2}{1 + (0.0366)(0.25)} \right\}$$
$$= 0.8476$$

Note that C_b can be determined directly from Fig. 5.5 if the coating material is PL-1. Finally, from Eq. (5.7),

$$C_f^\epsilon = \frac{2793}{(0.8476)(2)} = 1647.6 \ \mu\text{in./in. fringe}$$

For convenience, calibrators for photoelastic coatings with a standard calibration beam are provided by manufacturers. Figure 5.7a shows an example of a calibrator (Model 010-B) provided by the Micro-Measurements Group, Inc., and Fig. 5.7b illustrates the setup for normal-incident fringe order measurements for calibration of the coating. The deflections (W) at the loading point are read directly from the micrometer on the calibrator, and the corresponding fringe order (N) at a selected point on the coating is recorded. Thus the coating fringe value can be calculated by

$$C_f^\epsilon = \frac{\epsilon_1^s(1 + \mu^s)}{NC_b} = \frac{3d^s(1 + \mu^s)L_0}{2C_bL^3} \frac{W}{N} \tag{5.9}$$

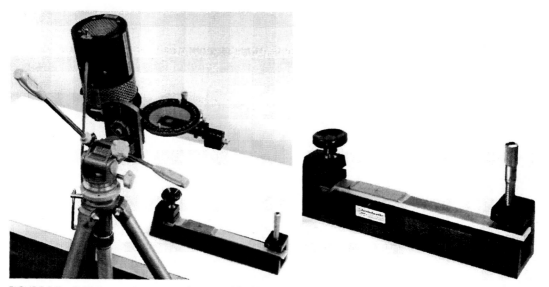

FIGURE 5.7   Calibrator and apparatus for normal-incidence fringe-order measurements [3].

where d^s and μ^s are the thickness and Poisson's ratio for the beam, C_b the correction factor for the beam in bending, L_0 the distance between the loading point and the measurement point, and L the length of the beam measured from the loading point and the fixed end. To average out the small errors in the measurements of W and N, the least squares method is used to draw a straight line on the W–N plot, and the slope of that line is used in Eq. (5.9) to replace the term W/N.

5.6   PRINCIPAL STRAIN SEPARATION METHODS

Theoretically, the stress separation methods described in Section 4.10 are equally valid for photoelastic coatings in two dimensions. However, when coatings apply to structural components, such as plates and shells, the shear-difference method, which is the most widely used method in photoelasticity, cannot be employed, since the shear stresses acting at the interface surface of the coating are unknowns. For this reason, the most widely used separation methods in photoelastic coating applications are the oblique-incidence method [12, 13], the strip coating method [4], and the strain gage separation method [5].

Oblique-Incidence Method

For illustration, consider the case where the directions of principal strain are known. For simplicity, assume that strains in coating and specimen are identical so that the superscripts in strains can be omitted in the following derivations. Assume also that the same coating thickness is used in the calibration as well as in the experiment. Therefore, for the normal-incidence method,

$$N = \frac{\epsilon_1 - \epsilon_2}{C_f^\epsilon} \tag{5.6a}$$

Let N_α be the fringe order obtained from an oblique-incidence method, which is achieved by rotating the direction of incident light about the ϵ_1 axis with an angle α as shown in Fig. 5.8. Since the light traverses a distance of $2(d^c/\cos\alpha)$ in direction 3', using Eq. (5.6),

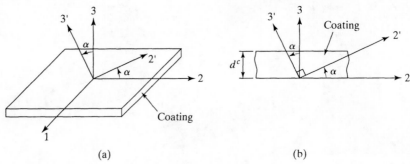

FIGURE 5.8   Coordinate system in oblique-incidence method.

$$N_\alpha = \frac{\epsilon_1 - \epsilon_2'}{C_f^\epsilon \cos \alpha} \tag{a}$$

where ϵ_2' is the secondary principal strain in the plane normal to the incident light. This can be found from equations of strain transformation and using Fig. 5.8(b), namely,

$$\epsilon_2' = \epsilon_2 \cos^2\alpha + \epsilon_3 \sin^2\alpha \tag{b}$$

The coating is under plane stress state, therefore

$$\epsilon_3 = \frac{-\mu^c(\epsilon_1 + \epsilon_2)}{1 - \mu^c} \tag{c}$$

Substituting Eqs. (c) and (b) into Eq. (a) gives

$$N_\alpha = \frac{\epsilon_1(1 - \mu^c \cos^2\alpha) - \epsilon_2(\cos^2\alpha - \mu^c)}{C_f^\epsilon (1 - \mu^c) \cos \alpha} \tag{5.10}$$

Solving Eqs. (5.6a) and (5.10) simultaneously gives

$$\epsilon_1 = \frac{C_f^\epsilon[N_\alpha(1 - \mu^c) \cos \alpha - N(\cos^2\alpha - \mu^c)]}{(1 + \mu^c) \sin^2\alpha}$$
$$\epsilon_2 = \frac{C_f^\epsilon[N_\alpha(1 - \mu^c) \cos \alpha - N(1 - \mu^c \cos^2\alpha)]}{(1 + \mu^c) \sin^2\alpha} \tag{5.11}$$

Hence the principal strains are determined.

Similarly, if N_β is the fringe order obtained from an oblique-incidence method, which is achieved by rotating the direction of incident light about the ϵ_2 axis with an angle β, we have

$$\epsilon_1 = \frac{C_f^\epsilon[N(1 - \mu^c \cos^2\beta) - N_\beta(1 - \mu^c) \cos \beta]}{(1 + \mu^c) \sin^2\beta}$$
$$\epsilon_2 = \frac{C_f^\epsilon[N(\cos^2\beta - \mu^c) - N_\beta(1 - \mu^c) \cos \beta)]}{(1 + \mu^c) \sin^2\beta} \tag{5.12}$$

In practice, to check the experimental results, the oblique-incidence experiments are performed twice by rotating the incident light about the ϵ_1 and ϵ_2 axes. The results, calculated by using Eqs. (5.11) and (5.12), should be the same.

When making measurements by using the oblique-incidence method, an oblique-incidence adapter (Model 033) with a fixed mirror angle is attached to the basic analyzer, as shown in Fig. 5.9a. The corresponding light path is shown in Fig. 5.9b. The light emerging from the polarizer is reflected by the oblique-incidence mirror (M), whose angle is fixed to simplify data reduction. The reflected light then traverses the photoelastic coating and is reflected back to the mirror; finally, the light is reflected by the mirror back to the analyzer.

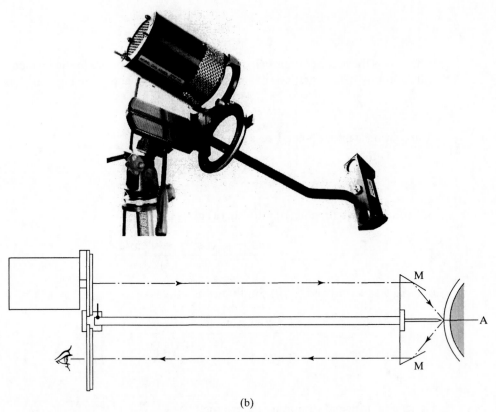

(b)

FIGURE 5.9 Instrument and light path in the oblique-incidence method [1].

Strip Coating Method

A strip coating consists of a large number of closely spaced parallel narrow strips (the thickness of the strip is several times the width), as shown in Fig. 5.10. When the strip coating is bonded on a specimen along a certain direction (e.g., x), the strips, similar to resistance strain gages, tend to transmit the strains (ϵ_x) along the x direction only. Thus

$$N_x \approx \frac{\epsilon_x}{C_f^\epsilon} \qquad (5.13)$$

From Eq. (5.13) it can be seen that the fringe order in the strip coating is approximately proportional to the strain along the strip direction. If the space between strips is very small, the regions of extinction in the strips blend together to form continuous fringes, as illustrated in Fig. 5.11. There are three different techniques that can be used to separate the principal strains by the strip coating method:

1. Similar to strain gage rosette applications, three strip coatings are used to determine the strains in three different directions; thus the principal strains and their directions can be calculated by Eq. (3.20).

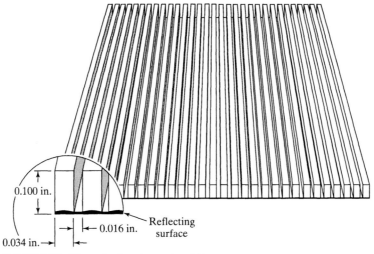

FIGURE 5.10   Photoelastic strip coatings [4].

2. Two strip coatings are used in any two orthogonal directions (say, the x and y directions) to determine the sum of the principal strains; and one continuous coating is used to determine the difference of the principal strains. From Eqs. (5.13) and (5.6) we have

$$\epsilon_1 + \epsilon_2 = \epsilon_x + \epsilon_y = C_f^\epsilon(N_x + N_y) \tag{d}$$

$$\epsilon_1 - \epsilon_2 = C_f^\epsilon N \tag{e}$$

where N is the fringe order obtained by using the continuous coating, and N_x and N_y are fringe orders obtained by using the strip coatings in the x and y directions, respectively. Solving Eqs. (d) and (e) simultaneously gives

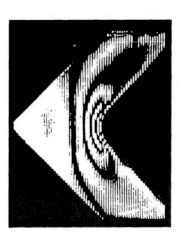

FIGURE 5.11   Fringe pattern obtained by the strip coating method for a knee frame under compression [4].

$$\epsilon_1 = \frac{C_f^\epsilon(N_x + N_y + N)}{2}$$
$$\epsilon_2 = \frac{C_f^\epsilon(N_x + N_y - N)}{2}$$

(5.14)

3. One strip coating is used in the x direction to measure strain ϵ_x; one continuous coating is used to measure the difference of the principal strains $(\epsilon_1 - \epsilon_2)$ from the isochromatic fringe patterns and to determine the principal direction β_1 from the isoclinic patterns. By using Eqs. (1.34), (5.6a), and (5.13), the following equations can be obtained:

$$\epsilon_1 = C_f^\epsilon(N_x + N \sin^2\beta)$$
$$\epsilon_2 = C_f^\epsilon(N_x - N \cos^2\beta)$$

(5.15)

where N and N_x are fringe orders obtained by continuous and strip coatings, respectively. Once the principal strains are found, the principal stresses of the specimen can be obtained by using Eq. (5.1b) and Hooke's law [Eq. (1.47)].

Strain Gage Separation Method

Electrical resistance strain gages can be used to provide additional information required to separate the principal strains. From the isoclinic and isochromatic fringe patterns of the coating, the principal direction β_1 and the difference of the principal strains $\epsilon_1 - \epsilon_2$ for the entire field are known. By bonding a single-element strain gage along the ϵ_1 direction at the points of interest, ϵ_1 can be determined by the strain gage reading ϵ_g. Thus

$$\epsilon_1 = \epsilon_g$$
$$\epsilon_2 = \epsilon_g - NC_f^\epsilon$$

(5.16)

where N is the fringe order. This method is conceptually simple; however, a number of practical problems occur, such as local reinforcement effects (dependent on the type of gage) and drift of strain readings due to the self-heating effect (the thermal conductivity of the coating is very low). The error introduced cannot readily be corrected.

To overcome the shortcomings mentioned above, special strain gages have been developed by the manufacturer, for example the PhotoStress separator gage provided by the Measurements Group, Inc. [1,5]. This gage is similar to the strain gage illustrated in Fig. 2.3m in shape, but it has several additional advantages. The gage alignment during bonding is not crucial since it gives the sum of principal strains $(\epsilon_1 + \epsilon_2)/2$ directly without any particular angular orientation. The residual reinforcement error can be eliminated by bonding the gage on standard PhotoStress coatings to calibrate for its effective gage factor. Soldering is not required since preattached lead wires are provided, and a protective coating is usually unnecessary because the gage grid is encapsulated in polyamide. To use commercial strain indicators with a power level low enough to avoid the self-heating effect, the resistance of the gage is 200 Ω, and the gage is connected to the strain indicator via a specially designed interface model [1,5]. When the gage reading is ϵ_g,

$$\epsilon_g = \frac{\epsilon_1 + \epsilon_2}{2}$$

Since

$$\epsilon_1 - \epsilon_2 = NC_f^{\epsilon}$$

$$\epsilon_1 = \epsilon_g + \frac{NC_f^{\epsilon}}{2} \tag{5.17}$$

$$\epsilon_2 = \epsilon_g - \frac{NC_f^{\epsilon}}{2}$$

EXAMPLE 5.2

A heavy steel specimen has been bonded with a photoelastic coating (type PL-1). The coating thickness is 0.075 in. and its fringe value is $1513\mu\epsilon$/fringe. Under loading, a fringe order of 2.5 is found at a point of interest on the coating by the normal-incidence method. The load is then removed and a PhotoStress separator gage is mounted on the coating at that location with an arbitrary gage orientation. The same load is reapplied, and it is found that the indicated strain is $300\mu\epsilon$. Determine the principal strains and stresses in the specimen. Young's modulus and Poisson's ratio of steel are 30,000 ksi and 0.29, respectively.

Solution   Since $C_f^{\epsilon} = 1513\mu\epsilon$/fringe, $\epsilon_g = 300\mu\epsilon$, and $N = 2.5$, by Eq. (5.17) we have

$$\epsilon_1 = 300 + \frac{(2.5)(1513)}{2} = 2191\mu\epsilon$$

$$\epsilon_2 = 300 - \frac{(2.5)(1513)}{2} = -1591\mu\epsilon$$

Using Hooke's law for the case of plane stress, the principal stresses can be found by

$$\sigma_1 = \frac{30 \times 10^6}{1 - 0.29^2}[2191 - (0.29)(1591)] \times 10^{-6} = 56.7\,\text{ksi}$$

$$\sigma_2 = \frac{30 \times 10^6}{1 - 0.29^2}[-1591 + (0.29)(2191)] \times 10^{-6} = -31.3\,\text{ksi}$$

In Example 5.2 the correction for reinforcing effects is not necessary, since it is negligible. However, as mentioned earlier in Section 5.4, when the coating is not thin or the modulus of the specimen material is relatively low, corrections must be made for both N and ϵ_g. Thus

$$\epsilon_g = C_i \epsilon_g' \tag{5.18}$$

where C_i is the correction factor and ϵ_g' is the indicated strain.

When the coated specimen is subjected to a biaxial stress loading, the correction factors for both N and ϵ_g are the same (i.e., C_p) and are given by Eq. (5.8a), since the strain at the coating surface, as measured by the gage, is the same as the average strain of the coating. However, when the coated plate is subjected to flexure, the correction factor for ϵ_g is slightly different than C_b as given by Eq. (5.8b), since the measured

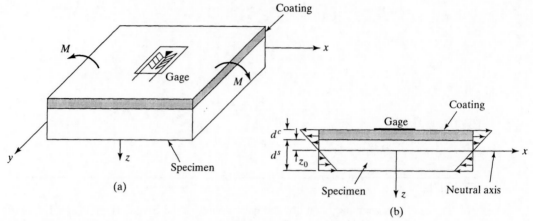

FIGURE 5.12   Coated element of a plate in pure bending.

strain by the gage is not equal to the average strain of the coating. The detailed derivation is as follows [2].

Consider a coated plate that is subjected to a pure bending moment, shown in Fig. 5.12a. Assume that the strain distribution is linear through the thickness, as shown in Fig. 5.12b. From elementary plate theory, we have

$$\epsilon_x^s = \frac{-z}{\rho} \qquad \text{for } -z_0 \le z \le (d^s - z_0)$$

$$\epsilon_x^c = \frac{-z}{\rho} \qquad \text{for } -(d^c + z_0) \le z \le -z_0 \qquad (5.19)$$

$$\epsilon_y^s = \epsilon_y^c = 0 \qquad \text{for all } z$$

where z is measured from the neutral axis (x axis) and z_0 is the distance between the neutral axis and the interface, shown in Fig. 5.12b. By using Hooke's law [Eq. (1.47)], the stresses can be expressed as

$$\sigma_x^s = -\frac{E^s}{1 - (\mu^s)^2} \frac{z}{\rho} \qquad \text{for } -z_0 \le z \le (d^s - z_0)$$

$$\sigma_x^c = -\frac{E^c}{1 - (\mu^c)^2} \frac{z}{\rho} \qquad \text{for } -(d^c + z_0) \le z \le -z_0 \qquad (5.20)$$

The position of the neutral axis, z_0, can be determined by the condition of force equilibrium in the x direction, namely,

$$\int_{-(d^c + z_0)}^{-z_0} \sigma_x^c \, dz + \int_{-z_0}^{d^s - z_0} \sigma_x^s \, dz = 0 \qquad (5.21)$$

Substituting Eq. (5.20) into Eq. (5.21), integrating, and solving for z_0 gives

$$z_0 = \frac{d^s(1 - \alpha\beta^2)/2}{1 + \alpha\beta} \qquad (5.22)$$

where α and β are as given in Eq. (5.8c).

The radius of curvature ρ can be obtained by the condition of moment equilibrium:

$$M = -\int_{-(d^c + z_0)}^{-z_0} \sigma_x^c z\, dz - \int_{-z_0}^{d^s - z_0} \sigma_x^s z\, dz = 0 \qquad (5.23)$$

Substituting Eq. (5.20) into Eq. (5.21), integrating, and solving for ρ gives

$$\frac{1}{\rho} = \frac{12M(1 - \mu^{s2})}{BE^s\,(d^s)^3} \qquad (5.24)$$

where

$$B = 4(1 + \alpha\beta^3) - \frac{3(1 - \alpha\beta^2)^2}{(1 + \alpha\beta)}$$

The strain measured by the PhotoStress separator gage is at the surface of the coating [i.e., $z = -(d^c + z_0)$]. Thus

$$\epsilon_x^c = -\frac{z}{\rho} = \frac{12M[1 - (\mu^s)^2](d^c + z_0)}{BE^s(d^s)^3} = \frac{6M[1 - (\mu^s)^2](1 + 2\beta + \alpha\beta^2)}{BE^s(d^s)^2(1 + \alpha\beta)} \qquad (e)$$

However, the strain at the surface of an uncoated plate ($z = -d^s/2$) is given by

$$\epsilon_x^s = -\frac{z}{\rho} = \frac{12M[1 - (\mu^s)^2]}{BE^s(d^s)^3} \frac{d^s}{2} = \frac{6M[1 - (\mu^s)^2]}{E^s(d^s)^2} \qquad (f)$$

Therefore, the correction factor for plates or beams in bending, denoted by C_{bg}, is given by

$$C_{bg} = \frac{\epsilon_x^s}{\epsilon_x^c} \qquad (g)$$

Substituting Eqs. (e) and (f) into Eq. (g) gives

$$C_{bg} = \frac{1 + \alpha\beta}{1 + 2\beta + \alpha\beta^2} \left[4(1 + \alpha\beta^3) - \frac{3(1 - \alpha\beta^2)^2}{1 + \alpha\beta} \right] \qquad (5.25)$$

where

$$\alpha = \frac{(1 - [\mu^s]^2)E^c}{(1 - [\mu^c]^2)E^s} \qquad \beta = \frac{d^c}{d^s} \qquad (5.8c)$$

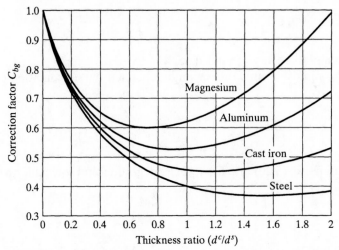

FIGURE 5.13   Correction factor C_{b_g} versus ratio of thickness in bending.

Equation (5.25) is shown in Fig. 5.13 for four specimen materials (steel, cast iron, aluminum, and magnesium) coated with the photoelastic material PL-1. It can be seen from Figs. 5.13 and 5.5 that C_{bg} is less than unity for all four materials when β is less than 2. This differs somewhat from C_b because the measurement is made at the surface of coating instead of at midthickness.

5.7   COATING SENSITIVITY AND SELECTIONS

Coating sensitivity (or coating fringe value) C_f^ϵ, defined by Eq. (5.6), represents the difference in principal strains (or the maximum shear strain γ_{max}) required to produce one fringe. It is perhaps one of the most important factors to be considered in the selection of a coating. From Eq. (5.6) it can be seen that the lower the value of C_f^ϵ, the more sensitive is the coating. Equation (5.6) can be rewritten as

$$C_f^\epsilon = \frac{\epsilon_1 - \epsilon_2}{N} = \frac{\gamma_{max}}{N} = \frac{\text{expected strain level}}{\text{desired maximum fringe number}} \qquad (5.26a)$$

$$= \frac{\lambda}{2d^c k} \qquad (5.26b)$$

where λ is the wavelength of light used in polariscope (22.7×10^{-6} in. or 0.577×10^{-6} m for white light), and k is the strain-optic coefficient of the coating, which is dimensionless and is provided by the manufacturer.

Ideally, the maximum expected strain level will be the yield strain for a coating applied to elastic stress problems. The expected maximum number of fringes to be measured depends on the type of instrumentation employed and the test conditions. For example, in a laboratory testing room with shades, the maximum number of fringes to be observed is four when a reflection polariscope (Model 031 [7]) with white light is used for a static test. In practice, for convenience in coating selection, Eq. (5.26b) is shown graphically in Fig. 5.14 for white light and different materials. Standard coating thicknesses and the ranges of k for available materials (for solid flat sheets only) are also shown in this figure.

EXAMPLE 5.3

Assume that the expected strain level is $4000\mu\epsilon$ and the maximum observable number of fringes is four for a coating illuminated with white light. Determine the coating sensitivity and find whether a standard thickness for a solid flat sheet is available from Fig. 5.14.

Solution By Eq. (5.26a), $C_f^\epsilon = 4000/4 = 1000\mu\epsilon$/fringe. In Fig. 5.14 we locate the point $1000\mu\epsilon$/fringe along the ordinate and then draw a line horizontally passing through that point and find that an intersection with a particular thickness line (0.080 in.) falls within the darked zone ($k = 0.10$ to 0.15). Therefore, a standard thickness of 0.080 in. for a flat sheet with a k of approximately 0.15 can be selected.

When a plate (e.g., a thin aluminum diaphragm) is subjected to an unknown combination of membrane and bending stresses, the coating thickness could be selected such that the correction factor C_b is equal to unity, thus leaving only a predictable maximum error for the plane stress reinforcing effect. This selection leads to a coating thickness greater than the plate thickness, and it is appropriate only when the plate is very thin. Alternatively, the coating thickness could be chosen so that the correction factors C_p and C_b are equal. A single common correction factor could then be applied directly to the fringe order measured.

EXAMPLE 5.4

A thin aluminum diaphragm with a thickness of 0.07 in. is subjected to an unknown combination of bending and membrane stresses. In order to have only a plane stress reinforcing effect, the coating (PL-1) thickness is selected so that $C_b = 1$. Find the coating thickness and the correction factor C_p.

Solution Since the coating is PL-1, from Fig. 5.5 it is found that the ratio of the thickness β is approximately 1.56. Thus

$$d^c = 1.56d^s = (1.56)(0.07) = 0.11 \text{ in.}$$

Also, from Fig. 5.4, the correction factor for $\beta = 1.56$ is given by $C_p = 1.064$.

In Examples 5.3 and 5.4, the selection of a solid flat sheet is demonstrated. When the test region of a structure is flat, it is preferable to use this type of coating since its thickness and its physical and photoelastic properties are uniform, its handling is easy,

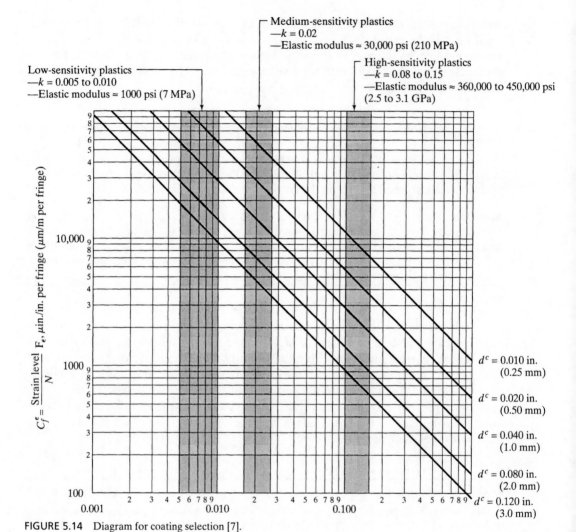

FIGURE 5.14   Diagram for coating selection [7].

and it is available from stock. In addition to solid flat sheets, two other basic forms are available for photoelastic coatings: liquids for casting contoured sheets and sprays for applications when curved surfaces associated with three-dimensional structural components or irregularly shaped structures are involved. A relatively uniform thickness of the plastic sheet can easily be achieved using the contoured-sheet method, so it is widely used in practice either to fabricate complex thin-section models for three-dimensional photoelastic stress analysis [8] or to coat directly on irregularly shaped structural components. On the other hand, liquids for spray application are limited to qualitative studies, since it is very difficult to obtain a relatively uniform coating thickness with spraying.

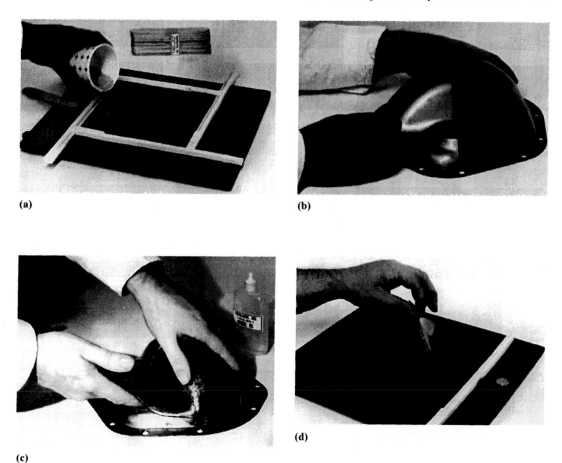

(a)

(b)

(c)

(d)

FIGURE 5.15 Major steps in the contoured-sheet method [8] (a) pouring the resin on a flat plate; (b) removing the semipolymerized plastic sheet; (c) contouring the plastic sheet to the desired shape; (d) removing the hardened plastic shell from the pattern.

5.8   COATING APPLICATIONS

For plane components, a precured flat sheet of coating is used. The sheet is machined to size before it is bonded to the test part. Alternatively, it can be machined to a proper shape after the coating is bonded to the specimen, and the specimen can be used as a template during machining. For irregularly shaped structures, the contoured-sheet method is used to obtain the proper-shaped coatings before bonding. The major steps of the contoured-sheet method are depicted in Fig. 5.15 and are described below.

First, mixed epoxy or resin (e.g., PLM-1) is poured on a flat plate, which is treated with a releasing agent (e.g., Teflon or silicone rubber) to cast a sheet with the desired thickness, as shown in Fig. 5.15a. During the semipolymerized stage (Fig. 5.15b), the plastic sheet is removed from the plate. The semipolymerized sheet is soft and pliable and its strain-optic coefficient is very low. It can therefore be contoured to the shape desired over a pattern without introducing photoelastic response, as shown in Fig. 5.15c. The contoured sheet is then allowed to fully polymerize, but it must remain in contact with the pattern during this stage. When the polymerization is complete, the hardened plastic shell is removed from the pattern, as shown in Fig. 5.15d. In this way, the required number of contoured plastic pieces have been made and are ready to be bonded to the specimen. The contoured method enables the application of coatings to specimens with almost any shape while maintaining the coating in a relatively stress-free state.

As with applications of electrical-resistance strain gages, the perfect bond between the coating and the test components is critical. Attention must be paid to surface preparations for both the coating and the specimen, as well as to the choice of the adhesive. The bonding procedures must be followed carefully. The surfaces are normally smoothed by sanding and cleaned by using suitable solvents to free them of all foreign materials.

The adhesive is normally an aluminum-filled polymer. To bond the coating, the adhesive is spread evenly over the surface of the test component. The shaped coating is then positioned at one end and slowly rotated and pressed into position to squeeze out any air bubbles. To avoid the introduction of residual stresses, pressure is normally not applied during the curing process. More details may be found in reference [8] for coating applications and in reference [9] for selecting photoelastic coatings (sheets and liquids) and adhesives.

PROBLEMS

5.1. Verify Eq. (5.3).

5.2. In applications of photoelastic coatings to elastic stress problems, the maximum fringe order occurs when the maximum shear stress on the specimen reaches its yield stress according to the Tresca yield criterion, namely, $(\sigma_1^s - \sigma_2^s)_{max} = \sigma_{ys}$. Derive an expression of the maximum fringe order (N_{max}) in terms of coating thickness (d^c), material strain fringe value (f_ϵ), and specimen material constants (i.e., Young's modulus E^s, Poisson's ratio μ^s, and yield stress in tension σ_{ys}).

5.3. A coating with a thickness of 0.06 in. and a material strain fringe value (M_f^ϵ) of 0.15×10^{-3} in./fringe is to be bonded to various specimens of different materials. Determine the maximum fringe orders (N_{max}) that could be developed for the following materials: mild steel AISI 1010 (E^s = 30,050 ksi, μ^s = 0.30, and σ_{ys} = 50.0 ksi), aluminum 2024-T3 (E^s = 10,000 ksi, μ^s = 0.33, and σ_{ys} = 25.4 ksi), red brass (E^s = 16,700 ksi, μ^s = 0.32, and σ_{ys} = 50.0 ksi), and magnesium M1 (E^s = 6500 ksi, μ^s = 0.35, and σ_{ys} = 24.7 ksi).

5.4. Figure P5.4 illustrates stress components in coated and uncoated elements from a specimen subjected to a plane stress loading. The reinforcing effect is corrected by a factor C_p, given by Eq. (5.8a). Verify Eq. (5.8a).

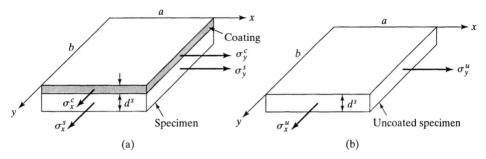

FIGURE P5.4   Stresses in coated and uncoated elements subjected to a biaxial stress loading.

5.5. Figure P5.5 illustrates strain distributions in a coated element taken from a beam subjected to pure bending ($\epsilon^s_y = \epsilon^c_y = 0$). The effects of strain variation through the thickness and of reinforcement from the coating are corrected by a correction factor C_b, given by Eq. (5.8b). Verify Eq. (5.8b). [*Hint:* First determine the position (z_0) of the neutral axis by considering force equilibrium in the x direction, then compute the radius of curvature (ρ) by considering moment equilibrium. Finally, obtain C_b by comparing the average strain in the coating with that of the uncoated specimen surface.]

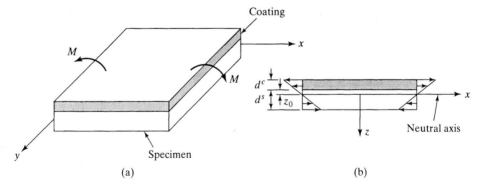

FIGURE P5.5   Stresses in coated and uncoated elements taken from a beam subjected to pure bending.

5.6. A coating (PL-1) with a thickness of 0.11 in. and a material strain fringe value (M^ϵ_f) of 0.15×10^{-3} in./fringe is to be bonded to an aluminum (E^s = 10,000 ksi and μ^s = 0.33) specimen with a thickness of 0.22 in., subjected to a tensile stress. Determine the fringe order (N) that could develop at the maximum stress σ_1 = 20 ksi with and without taking into account the reinforcing effect.

5.7. Determine the correction factors C_p and C_b for an aluminum specimen for thickness ratios (d^c/d^s) of 0.1, 0.2, 0.5, 1.0, 1.5, and 2.0. Young's modulus and Poisson's ratio for the coating are 2.90 GPa and 0.36, respectively.

5.8. To determine the coating fringe value (C_f^ϵ), a coating with a thickness of d^c is bonded to a tension strip of known dimensions w and d^s. At load P, the fringe order N is measured with a reflective polariscope. Derive an expression for C_f^ϵ of the coating. Young's moduli and Poisson's ratios for specimen and coating are E^s, E^c, μ^s, and μ^c, respectively.

5.9. A heavy steel specimen (E^s= 30,000 ksi and μ^s = 0.29) has been coated with photoelastic coating (μ^c = 0.36 and C_f^ϵ = 750$\mu\epsilon$/fringe). At a critical point, a fringe order of 3.1 is measured by using the normal-incidence method, and a fringe order of 2.2 is measured by using the oblique-incidence method and rotating by 35° the incident light about the ϵ_2 axis. Determine the principal stresses at the critical point on the surface of the specimen.

5.10. Verify Eq. (5.14).

5.11. To determine the principal stress distribution of an aluminum knee frame, which is subjected to a pair of concentrated loads shown in Fig. P5.11, a continuous coating and two strip coatings (along the x and y axes) are applied to the frame successively. At point A, the corresponding fringe orders are recorded as 3.1, 2.0, and 0.5. Calculate the principal stresses at point A. C_f^ϵ = 750$\mu\epsilon$/fringe.

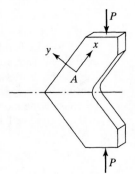

FIGURE P5.11   Aluminum knee frame.

5.12. A thin aluminum tensile specimen (0.13 in. thick) has been coated with PL-1 (0.09 in. thick), whose fringe value C_f^ϵ is 1261$\mu\epsilon$/fringe. At a tensile loading, the uncorrected fringe order at a point of interest is 1.7. The specimen is then unloaded, a PhotoStress separator gage is installed on the coating at the same point, and the load is reapplied. An uncorrected strain of 850$\mu\epsilon$ is recorded. Determine the principal stresses at the point of interest with and without correction for reinforcing effects. What is the error if correction is not made?

5.13 The same specimen as in Problem 5.11 is subjected to pure bending instead of an in-plane force. All other parameters remain the same. Determine the principal stresses at the point of interest and the error if correction is not made.

5.14. The expected strain level for an aluminum specimen is 2700$\mu\epsilon$ and the maximum number of desired fringe order is 3. Determine the coating thickness from Fig. 5.14.

5.15. A thin aluminum diaphragm with a thickness of 0.07 in. is subjected to an unknown combination of membrane and bending stresses. Determine the coating thickness (E^c = 420 ksi and μ^c = 0.36) necessary so that only a single common correction factor is used to the measured fringe order, namely, $C_p = C_b$. What is the correction factor?

References

[1] *Introduction to Stress Analysis by the PhotoStress Method,* Tech Notes TN-702 and TN-702-1, Measurements Group, Inc., Raleigh (North Carolina), 1980 and 1989.

[2] F. Zandman, S. S. Redner, and E. I. Riegner, Reinforcing effect of birefringent coating, *Exp. Mech.,* **2,** 55–64, 1962.

[3] *Calibration of Photoelastic Coatings,* Tech Note TN-701, Measurements Group, Inc., Raleigh (North Carolina), 1977.

[4] R. O'Regan, New method for determining strain on the surface of a body with photoelastic coatings, *Exp. Mech.,* **5,** 241–246, 1965.

[5] *Principal Stress Separation in PhotoStress Measurements,* Tech Note TN-708, Measurements Group, Inc., Raleigh (North Carolina), 1986.

[7] *How to Select Photoelastic Coatings,* Tech Note TN-704, Measurements Group, Inc., Raleigh (North Carolina), 1978.

[8] *Model Making For Three-Dimensional Photoelastic Stress Analysis,* Tech Note TN-707, Measurements Group, Inc., Raleigh (North Carolina), 1982.

[9] *Material for Photoelastic Coatings/Models,* Photolastic Bull. S-116-D, Measurements Group, Inc., Raleigh (North Carolina), 1983.

[10] F. Zandman, S. Redner, and J. W. Dally, *Photoelastic Coatings,* SESA Monograph 3, Iowa State University Press, Ames, Iowa, 1977.

[11] *Student Manual on the Photoelastic Coating Technique,* Bull. 315-A, Measurements Group, Inc., Raleigh (North Carolina), 1984.

[12] J. W. Dally and W. F. Riley, *Experimental Stress Analysis;* second edition, McGraw-Hill Publishing Co., 1978.

[13] S. S. Redner, New oblique incidence method for direct photoelastic measurement of principal strains, *Exp. Mech.,* **3,** 67–72, 1963.

CHAPTER 6

Geometric Moiré Techniques in Strain Analysis

6.1 INTRODUCTION

Moiré is generally used in engineering to describe interference fringes produced by superimposing two (or more) sets of gratings. The historical background for applications of moiré methods may be found in reference [1]. All moiré methods use moiré as the gage element. In this chapter we discuss only methods used to produce moiré fringes by mechanical means (i.e., geometric moiré).

Moiré fringe patterns can be used to measure various quantities, such as in-plane and out-of-plane displacements, rotations, strains, and curvatures. Since it is a geometric interference method, the moiré method can be applied to large deformation measurements, dynamic displacement or strain measurements, elastic–plastic strain measurements, and so on. Its disadvantage is that it is not sensitive enough to measure very small strains accurately. However, with the development of high-sensitivity moiré interferometry (the combination of the concepts and techniques of geometrical moiré and optical interferometry), this inadequacy can be compensated for and the moiré method is used more widely in experimental stress analysis. Interested readers are referred to reference [2] for details on moiré interferometry.

6.2 FUNDAMENTAL PROPERTY OF MOIRÉ FRINGES

Gratings are arrays of alternately opaque and transparent (or dark and light) lines which are usually straight and equally spaced, as shown in Fig. 6.1. There are some other gratings, such as concentric equally spaced circles, shown in Fig. 6.2a, and cross-gratings or dots made of two sets of equally spaced straight lines perpendicular to each other, shown in Fig. 6.2b and c. Unless otherwise stated, the gratings referred to in this

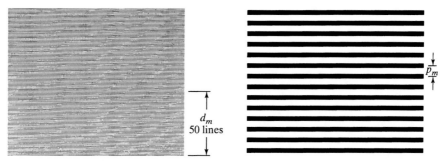

FIGURE 6.1   Typical gratings of equally spaced straight lines [1].

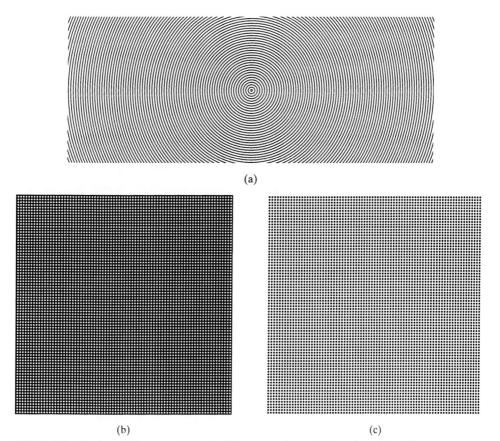

(a)

(b)   (c)

FIGURE 6.2   Grating of (a) concentric circles, (b) cross-grating, and (c) grating of dots [1].

chapter are equally spaced straight lines (Fig. 6.1). The ratio between the line width and spacing is 1.

Two gratings are necessary to produce moiré fringes. In the experimental measurement, one is called the *specimen* (or *model*) grating and the other is known as the *master* (or *reference*) grating. The center-to-center distance between the master grating lines is called *pitch,* designated by the symbol p_m; the number of lines per unit length for the master grating is known as the *density* of the master and represented by d_m, as shown in Fig. 6.1. The corresponding pitch and density of the specimen, which are changing during the deformation, are denoted by p_s and d_s, respectively. The direction perpendicular to the lines of the master grating is called the *primary direction* and the direction parallel to the lines of the master grating is known as the *secondary direction.*

Consider first the interference of two gratings with lines parallel to each other but of different pitches p_m and p_s, as shown in Fig. 6.3. This can be visualized as the case when one of them is the master grating and the other is the specimen grating, which is under uniform deformation. A light fringe will be observed at the place where two opaque strips coincide (point A' in Fig. 6.3) and the neighboring transparent area is a maximum, so that the *average* transmitted light intensity is also a maximum. Similarly, a dark fringe will be observed at the points where a transparent strip is covered by an opaque strip; in this case the average transmitted light is minimum, or almost zero as there is minimum transmission from the neighbouring areas. The fringe direction coincides with the secondary direction. Let δ be the center-to-center distance between the light (or dark) fringes (known as the *fringe spacing*) and m be the number of lines of the master grating within two light fringes. Thus we have

$$\delta = mp_m = (m \pm 1)p_s \qquad (6.1)$$

If the pitches of the master grating and the specimen grating were originally the same, the difference of the pitches is produced purely by the uniform deformation (either contraction or elongation) of the specimen. If the master grating and the specimen grating were initially coincident and the specimen grating is fixed at the left end, as shown in Fig. 6.3, a point A in the undeformed state has displaced an amount p and is now in A' at the centerline of a light fringe (called the *first light fringe*). The engineering or nominal strain, ϵ, can be computed by

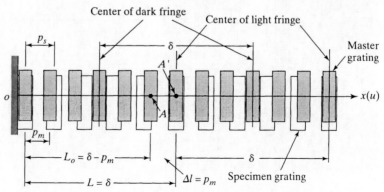

FIGURE 6.3   Formation of moiré fringes due to a difference in pitch [1].

$$\epsilon = \frac{\Delta L}{L_0} = \frac{p_m}{\delta - p_m} \simeq \frac{p_m}{\delta} \tag{6.2}$$

where p_m is known and δ can be directly measured from the moiré fringe patterns.

Similarly, at the centerline of the second light fringe, the displacement is $2p_m$, while at the centerline of the mth light fringe, the displacement is mp_m. In other words, moiré fringes are the loci of points with equal relative displacement u in the primary direction. Let m be the number of moiré fringes within the gage length G_l:

$$\epsilon = \begin{cases} \dfrac{\Delta L}{L_0} = \dfrac{mp_m}{G_l - mp_m} & \text{for tensile strain} \tag{6.3} \\[3mm] -\dfrac{\Delta L}{L_0} = -\dfrac{mp_m}{G_l + mp_m} & \text{for compressive strain} \tag{6.4} \end{cases}$$

Consider next the interference of two gratings by pure rotation (with no elongation or contraction). Both gratings have pitch p_m and one grating rotates an angle β counterclockwise with respect to the other grating, as shown in Fig. 6.4a. The line connecting the intersecting points is the centerline of the light fringe, since the average light intensity is maximum there. Figure 6.4b shows the moiré fringes formed by the pure rotation, where both gratings have a line density of 59 lines per inch (or 2.32 lines per millimeter). Note that the moiré fringes bisect the angle α ($= \pi - \beta$) between the lines of the two gratings.

From the geometry shown in Fig. 6.4a, we have

$$\alpha = \frac{\pi}{2} \pm \frac{\beta}{2} \tag{6.5a}$$

or

$$\alpha = \pm(2\alpha - \pi) \tag{6.5b}$$

where the plus sign applies for the case where both α and β are measured in the same direction from the master grating. Furthermore, the length of side a_1 is given by

$$a_1 = \frac{p_m}{\sin \beta}$$

and δ is given by

$$\delta = a_1 \cos \frac{\beta}{2} = \frac{p_m \cos(\beta/2)}{\sin \beta} = \frac{p_m}{2 \sin(\beta/2)}$$

Thus for a small angle β we have

$$\beta \simeq \frac{p_m}{\delta} \tag{6.6}$$

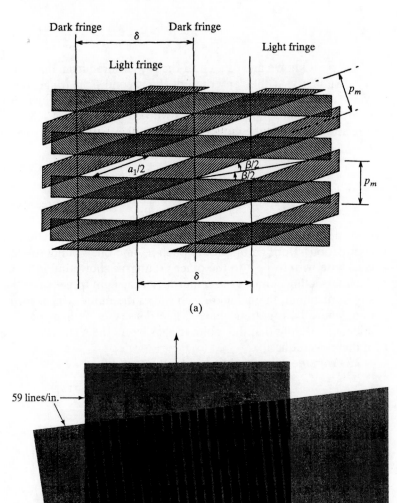

(a)

(b)

FIGURE 6.4 Moiré fringes formed by pure rotation [12].

6.3 GEOMETRICAL METHOD OF MOIRÉ FRINGE ANALYSIS IN TWO DIMENSIONS

Consider now the interference of two gratings with different pitches and a relative rotation with respect to each other [12], i.e. the combination of the two cases analyzed in Section 6.2, as shown in Fig. 6.5. The difference in pitches can be visualized as the result of elongation or contraction in the direction perpendicular to the lines of specimen grating, and the rotation can be attributed to shear deformation. These two effects occur simultaneously at a point in a stressed specimen under general loadings. In order

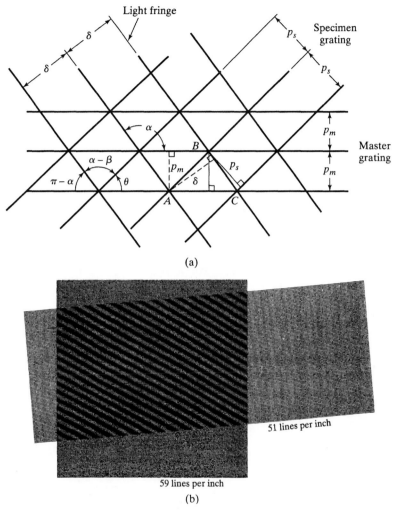

(a)

(b)

FIGURE 6.5   Moiré fringes formed by a combination of different pitches and rotations [1,12].

to simplify the plots, light fringes and opaque bars of gratings are represented by their centerlines, as shown in Fig. 6.5a.

Let's first determine the angle β, which is closely related to the shear strain. From the geometry shown in Fig. 6.5a, the segment AB gives

$$\frac{p_m}{\sin \beta} = \frac{\delta}{\sin (\alpha - \beta)} \qquad (6.7)$$

Solving for angle β gives

$$\beta = \tan^{-1} \frac{\sin \alpha}{\delta/p_m + \cos \alpha} \qquad (6.8)$$

Note that when rotation β is very small, α approaches $\pi/2$ as p_s approaches p_m. Therefore, Eq. (6.8) reduces to Eq. (6.6). Similarly, by considering the segment BC,

$$\frac{p_s}{\sin(\alpha - \beta)} = \frac{p}{\sin(\pi - \alpha)} = \frac{p}{\sin\alpha} \tag{6.9}$$

Using Eq. (6.7), we have

$$p_s = \frac{p_m \sin(\alpha - \beta)}{\sin\alpha} = \frac{\delta \sin\beta}{\sin\alpha} \tag{6.10}$$

Using Eq. (6.8) to eliminate θ gives

$$p_s = \frac{\delta}{\sqrt{1 + (\delta/p_m)^2 + 2(\delta/p_m)\cos\alpha}} \tag{6.11}$$

Thus p_s can be computed by using Eq. (6.11), since p_m is known, and δ and α can be measured from the moiré fringe patterns. In practice, the centerline of light fringes is easier to locate than those of dark fringes. From Eq. (6.10) it can be seen that α is 0 or π when β is zero. For many applications, rotation β is very small so that α approaches 0 or π. Equation (6.11) is simplified as

$$p_s = \frac{p_m \delta}{|p_m \pm \delta|} \tag{6.12}$$

It should be emphasized that Eqs. (6.6) and (6.12) cannot be used simultaneously even in the case when rotation β is very small, since different approximations were made in deriving these two equations. Once the deformed specimen pitch p_s is known, the engineering or nominal strain, ϵ, can be found as

$$\epsilon = \frac{p_s - p_m}{p_m} \tag{6.13}$$

EXAMPLE 6.1

The moiré fringe pattern shown in Fig. 6.5b can be visualized as having been produced by uniform elongation in the primary direction and by a shear deformation if both gratings had the same pitch and were initially coincident. Determine the engineering strain ϵ.

Solution   By direct measurement, $\alpha \approx 155°$ and $\delta \approx 0.115$ in.. Since $p_m = 1/59$ in., the deformed pitch p_s can be calculated by using Eq. (6.11),

$$p_s = \frac{0.115}{\sqrt{1 + [(59)(0.115)]^2 + 2[(59)(0.115)]\cos 155°}} = 0.0195$$

The engineering strain can be found by Eq. (6.13),

$$\epsilon = \frac{0.0195 - 1/59}{1/59} = 0.151$$

Since the actual deformed pitch is known (i.e., $p_s = 1/51$ in.), the error introduced in the measurement can be found as

$$\text{error} = \frac{1/51 - 0.0195}{1/51}(100) = 0.55\%$$

For this simple case, the measurement is quite accurate. Note that accuracy is commensurate with that of the measurements of α and δ.

As mentioned earlier, a typical two-dimensional homogeneous strain field exists if one grating with pitch p_m is considered as the master grating and is fixed and the other is considered as the specimen grating that coincided with the master grating originally and had a pitch p_m but was deformed to a pitch p_s and rotated at angle θ. Note that the strain computed by Eq. (6.13) is only an average strain between two fringes. Thus the geometrical moiré fringe analysis above should be limited to homogeneous strain fields or to very small regions in the nonhomogeneous fields.

In two-dimensional strain fields, there are three in-plane strain components: ϵ_x, ϵ_y, and γ_{xy}. In general, some further analysis with other gratings may be required to determine these strain components completely. An illustration follows. For simplicity, assume that ϵ_x, ϵ_y, and γ_{xy} are small and that ϵ_x and ϵ_y are both positive strains. First, both master grating and specimen grating with the same pitches are aligned in the x direction so that they are initially parallel to each other. After deformation, the positions of the two gratings are shown schematically, as in Fig. 6.6a (similar to Fig. 6.5a). Because rotation θ_y is very small, it moves the specimen grating lines only in the x direction, and because ϵ_x only elongates or contracts the specimen grating lines, they have no effect on the moiré fringes. Once δ_1 and ϕ_1 are obtained from the moiré fringe patterns and p is determined, by using Eq.(6.8), we have

$$\theta_x = \tan^{-1}\frac{\sin\phi_1}{\delta_1/p_m + \cos\phi_1} \tag{6.14}$$

Using Eq. (6.11), the deformed specimen pitch P_y' can be calculated by

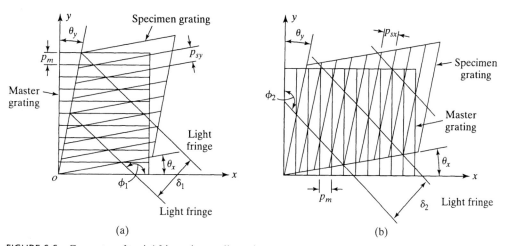

(a)   (b)

FIGURE 6.6   Geometry of moiré fringes in two dimensions.

$$p_{sy} = \frac{\delta}{\sqrt{1 + (\delta_1/p_m)^2 + 2(\delta_1/p_m)\cos\phi_1}} \tag{6.15}$$

Thus ϵ_y can be computed by using Eq. (6.13). Similarly, from Fig. 6.6b, θ_y and p_{sx} can be determined as

$$\theta_y = \tan^{-1}\frac{\sin\phi_2}{\delta_2/p_m + \cos\phi_2} \tag{6.16a}$$

$$p_{sx} = \frac{\delta_2}{\sqrt{1 + (\delta_2/p_m)^2 + 2(\delta_2/p_m)\cos\phi_2}} \tag{6.16b}$$

and ϵ_x can also be computed by using Eq. (6.13). In summary, two-dimensional strain components are

$$\epsilon_x = \frac{p_{sx} - p_m}{p_m} \qquad \epsilon_y = \frac{p_{sy} - p_m}{p_m} \qquad \gamma_{xy} = \theta_x + \theta_y \tag{6.17}$$

6.4 DISPLACEMENT METHOD OF MOIRÉ FRINGE ANALYSIS IN TWO DIMENSIONS

As mentioned in Section 6.4, since Moiré fringes are loci of points with equal displacement in the primary direction, they can be interpreted by a displacement field. For identification purposes, let the centerline of opaque bars of the master grating be numbered from 0 to L (or 0 to $-L$); those of the specimen grating can be numbered from 0 to M (or 0 to $-M$), and the centerline of light fringes can be numbered from 0 to N, as shown in Fig. 6.7 for a general two-dimensional deformation. It is easy to see from Fig. 6.7 that

$$N = L - M \tag{6.18a}$$

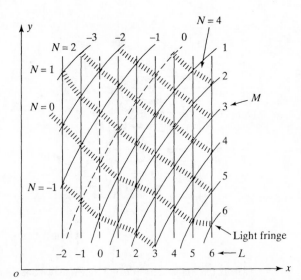

FIGURE 6.7 Moiré fringes in a general two-dimensional deformation [3].

These fringes, sometimes referred to as *subtractive fringes,* are those observed most frequently. However, it should be noted that in some instances there exist additive fringes that tend to obscure the information represented by the subtractive fringes [3]. The additive fringes are governed by the equation

$$N' = L + M \tag{6.18b}$$

In Fig. 6.7, assume that both the master grating and the specimen grating initially had a pitch p_m and were initially coincident along the y direction. The moiré fringes are assumed to be the subtractive ones governed by Eq. (6.18a). In general, displacement components in both the x and y directions, denoted by u and v, exist at points on the specimen. In Fig. 6.7, u, the displacement component in the primary direction, is related to the fringes, but v is indeterminate since deformation in the y direction does not affect moiré fringe patterns. It is easy to see from Fig. 6.7 that the zero-order moiré fringe is formed by the points of the specimen grating whose displacement component in the primary direction (u) is zero; the first-order fringe is formed by the points of the specimen grating whose displacement component (u) in the primary direction is p_m, and so on.

Similarly, in order to measure the displacement component v, it is necessary to orient both the master grating and specimen grating along the x direction. In practice, cross gratings are used for both the master grating and specimen grating [3]. With these two moiré fringe patterns, the strain field in the specimen can be determined.

EXAMPLE 6.2

A plexiglass beam is subjected to four-point loadings, as shown in Fig. E6.2a. The beam is 12 in. long, 0.25 in. thick, and 1.0 in. high. Both gratings had an initial pitch of 1/300 in. and the grating lines were in x direction. The master grating is fixed in space, and no fringes appeared at zero load. The moiré fringe pattern under load is shown in Fig. E6.2b. Determine the deflection curve along the centerline of the beam.

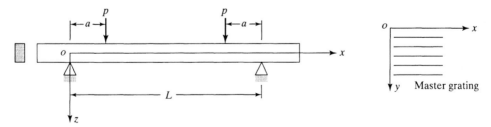

FIGURE E6.2a   Beam subjected to four-point loading.

FIGURE E6.2b   Moiré fringe pattern of a beam in pure bending [1].

Solution As we know, the moiré fringes are loci of points with equal displacement in the primary direction (z direction). At supports, there is no deflection, so the fringes are zero-order fringes. The remainder can readily be counted. Note that one fringe order corresponds to 1/300 in. of deflection; therefore, the deflection curve along the centerline of the beam can be obtained and plotted as in Fig. E6.2c.

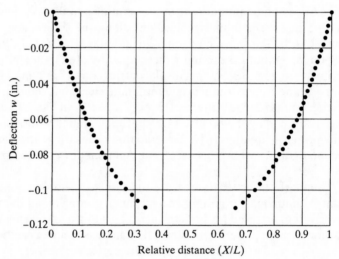

FIGURE E6.2c Deflection versus relative distance.

The general procedures to determine the strain field by the displacement method involve several steps. Suppose that the moiré fringe patterns are obtained for a two-dimensional model subjected to a general in-plane loading, as shown in Fig. 6.8. Since the primary direction is the x direction, the fringes are related to the displacement u. To determine the strain at point o, draw two straight lines (AB and CD) which pass through point o and parallel to the x and y axes, respectively. Curves of displacement u versus position along AB and CD can be plotted, as illustrated in Fig. 6.8, and the slopes ($\partial u/\partial x$ and $\partial u/\partial y$) at point o can then be determined. Using procedures similar to those for analyzing moiré fringes obtained by placing both the master grating and specimen grating initially along the x direction, $\partial v/\partial x$ and $\partial v/\partial y$ at point o can be obtained. Therefore, the strain components at point o for small strains can be completely determined once the four partial derivatives are obtained, namely,

$$\epsilon_x = \frac{\partial u}{\partial x}$$

$$\epsilon_y = \frac{\partial v}{\partial y} \tag{1.31}$$

$$\gamma_{xy} = \frac{\partial u}{\partial y} + \frac{\partial v}{\partial x}$$

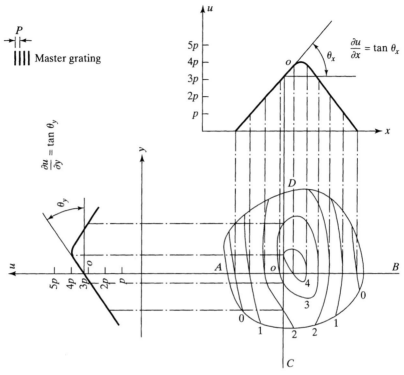

FIGURE 6.8   Displacement–position curve for determination of $\partial u/\partial x$ and $\partial u/\partial y$ at point o [3].

It should be noted that the method outlined above is not limited to small deformations. The method can be applied to large deformations, too. For large strains, high-order derivatives will be included in strain–displacement relationships. Furthermore, there are usually two different descriptions, the Lagrangian description and the Eulerian description. In the *Lagrangian description,* the strain components in two dimensions are defined as

$$\epsilon_x^L = \sqrt{1 + 2\frac{\partial u}{\partial x} + \left(\frac{\partial u}{\partial x}\right)^2 + \left(\frac{\partial v}{\partial x}\right)^2} - 1$$

$$\epsilon_y^L = \sqrt{1 + 2\frac{\partial v}{\partial y} + \left(\frac{\partial v}{\partial y}\right)^2 + \left(\frac{\partial u}{\partial y}\right)^2} - 1 \qquad (6.19)$$

$$\gamma_{xy}^L = \arcsin \frac{\partial u/\partial y + \partial v/\partial x + \partial u/\partial x\, \partial u/\partial y + \partial v/\partial x\, \partial v/\partial y}{(1 + \epsilon_x^L)(1 + \epsilon_y^L)}$$

In the *Eulerian description,* the strain components in two dimensions are given by

$$\epsilon_x^E = 1 - \sqrt{1 - 2\frac{\partial u}{\partial x} + \left(\frac{\partial u}{\partial x}\right)^2 + \left(\frac{\partial v}{\partial x}\right)^2}$$

$$\epsilon_y^E = 1 - \sqrt{1 - 2\frac{\partial v}{\partial y} + \left(\frac{\partial u}{\partial y}\right)^2 + \left(\frac{\partial v}{\partial y}\right)^2} \tag{6.20}$$

$$\gamma_{xy}^E = \arcsin \frac{\partial u/\partial y + \partial v/\partial x - \partial u/\partial x\, \partial u/\partial y - \partial v/\partial x\, \partial v/\partial y}{(1 - \epsilon_x^E)(1 - \epsilon_y^E)}$$

It is obvious that these procedures are rather tedious and time consuming for manually determining strains at a large number of points. The entire process of data analysis can be computerized if the fringes are recorded in digital form by a pattern-recognition technique [3] so that the displacement curves and the displacement derivatives can be computed by the cubic-spline method.

The accuracy of the strains determined by the displacement method depends largely on the accuracy of the displacement derivatives determined, which, in turn, depend on the numbers of fringes per unit length. In practice, the accuracy of direct derivatives $\partial u/\partial x$ and $\partial v/\partial y$ is generally acceptable. That is not the case for the cross derivatives $\partial u/\partial y$ and $\partial v/\partial x$. Slight errors in alignment of gratings (either master grating or specimen grating) with the x or y axes produce a combination of load-induced fringes and misaligned rotation fringes. In a rotation fringe pattern, the displacement gradient (direct derivative) in the primary direction is small, but the displacement gradient (cross derivative) is large in the direction parallel to the master grating lines (the secondary direction). To determine the shear strain more accurately, cross gratings for both master and specimen gratings are used to obtain u and v displacement fields. The errors introduced in the cross derivatives by rotation misalignment will be equal in magnitude but opposite in sign; thus their contributions to the shear strain are canceled out. In practice, the pitch of master and specimen gratings is constructed slightly differently to resolve the difficulty of existing interweaving between the two families of fringes. Figure 6.9 shows an example of moiré fringes obtained by crossed gratings with different pitches.

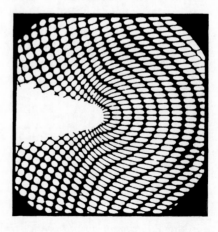

FIGURE 6.9   Moiré fringe obtained by cross gratings with different pitches [4].

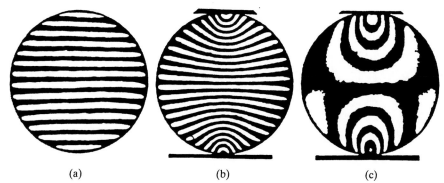

(a)   (b)   (c)

FIGURE 6.10   Moiré fringes by the linear mismatch method (a) M; (b) $M + L$; (c) L [3].

When the initial pitches of master and specimen gratings differ slightly, a mismatch fringe pattern known due to the *mismatch method* results. This can be used to increase the accuracy of strain measurements, since the fringe numbers are increased so that the data points for plotting the displacement curves will increase. There are usually three kinds of mismatch: linear mismatch resulting from two gratings with initially different pitches; rotational mismatch, which is due to two gratings with synonymous pitches but with different initial orientation; and a combination of linear and rotational mismatches. Figure 6.10 illustrates moiré fringes by a linear mismatch method, where M represents the linear mismatch, L represents the load, and $M + L$ represents the linear mismatch plus load.

In previous discussions, fringe orders were assumed to be already known. In practice, how to number the fringes in proper order is a rather challenging task, since fringe patterns alone are insufficient to determine proper fringe orders. For example, we cannot even determine if a moiré fringe pattern results from a homogeneous tensile strain field or from a homogeneous compressive strain field. In simple cases this difficulty can be overcome by using boundary conditions. For example, at points of zero displacement, the fringe order is zero. Once a zero-order fringe is recognized, the remainder can be numbered by direct counting either in increasing or decreasing order, since the displacement components are continuous, single-valued functions for a simply connected body. However, it is very important to determine whether it should be numbered in increasing order or in decreasing order (i.e., whether the derivatives are positive or negative), since that will result in a totally different displacement or strain field. In simple cases, as illustrated in Example 6.3, this is easy to do by noting the elastic stress or strain field.

EXAMPLE 6.3

A uniform beam is subjected to a pure bending moment. The moiré fringe pattern is shown in Fig. E6.3, where the primary direction coincides with the x axis (the neutral axis). Determine the proper number of the fringe orders.

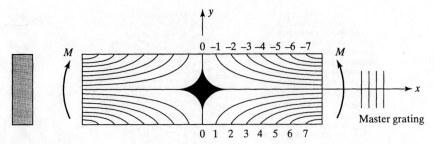

FIGURE E6.3

Solution   Since the beam is under pure bending, there is no elongation along the neutral axis. In other words, u is zero along the x axis, so the corresponding fringe is the zero-order fringe. From elementary beam theory, the upper portion of the beam is under compression, while the lower portion is under tension. Therefore, in the right hand half portion of the beam the fringe orders are negative in the upper portion, which are in decreasing order along the positive y direction; and the fringe orders are positive in the lower portion, which are in increasing order along the positive y direction. The complete numbering of fringes is also shown in Fig. E6.3.

In more general cases it is common practice to number moiré fringes in relative order, since only the derivatives of displacement are important to determine the strain components. To number the fringes properly without information on boundary conditions, we can use the mismatch method to determine whether the number of fringes are increasing or decreasing. Using Eq. (6.1), the fringe spacing is given by

or
$$\delta = \begin{cases} \dfrac{p_m p_s}{p_m - p_s} & \text{if } p_m > p_s \qquad (6.21a) \\[3mm] \dfrac{p_m p_s}{p_m - p_s} & \text{if } p_m < p_s \qquad (6.21b) \end{cases}$$

In the linear mismatch method, the pitch of the master grating is changed to $p_m + dp_m$ ($dp > 0$). The resulting fringe spacing becomes

$$\delta^* = \begin{cases} \delta + d\delta = \dfrac{(p_m + dp_m)\,p_s}{(p_m + dp_m) - p_s} \simeq \dfrac{p_m p_s}{(p_m - p_s) + dp_m} & \text{if } p_m > p_s \;\;(6.22a) \\[3mm] \delta + d\delta = \dfrac{(p_m + dp)\,p_s}{p_s - (p_m + dp_m)} \simeq \dfrac{p_m p_s}{(p_s - p_m) - dp_m} & \text{if } p_m < p_s \;\;(6.22b) \end{cases}$$

By comparing Eqs. (6.22) with Eqs. (6.21), we have

and
$$\begin{aligned} \delta^* &< \delta \qquad \text{when } p_m > p_s \\ \delta^* &> \delta \qquad \text{when } p_m < p_s \end{aligned} \qquad (6.23)$$

In other words, if the fringe spacing δ^* (due to an increase in master grating pitch dp_m) is smaller than the fringe spacing δ ($d\delta < 0$), as shown in Fig. 6.10, we can determine by using Eq. (6.22b) that the deformed specimen pitch p_s is smaller than p_m (i.e., the normal strain in the primary direction is compressive). The fringe order would then be numbered in decreasing order along the positive coordinate axis in the primary direction. If the fringe spacing δ^* (due to an increase of master grating pitch dp_m) is larger than the fringe spacing δ ($d\delta > 0$), we know that the deformed specimen pitch p_s is larger than p_m (i.e., the normal strain in the primary direction is a tensile strain so the fringe order should be numbered in increasing order along the positive coordinate axis in the primary direction). This numbering procedure can be illustrated further by the following example.

EXAMPLE 6.4

The moiré fringe pattern of a circular disk under a three-point loading is shown in Fig. E6.4a. A composite picture of the moiré fringe pattern with mismatch is shown in Fig. E6.4b, where the left half is obtained by imposing a positive dp_m to the master grating, and the right half is exactly the same as that in Fig. E6.4a. Determine the relative fringe orders.

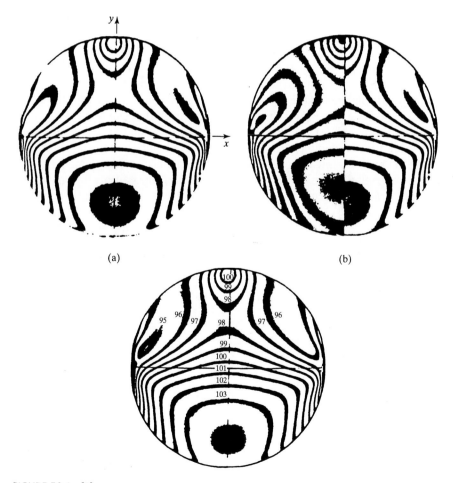

(a)   (b)

FIGURE E6.4   [3]

Solution   From Fig. E6.4a it can be seen that $\delta^* < \delta$ for the upper one-third of the fringes along the y axis, and $\delta^* > \delta$ for the lower two-thirds of the fringes along the y axis. Therefore, from Eq. (6.23), $p_m > p_s$ for the upper three fringes along the y axis in the composite picture on the right side, so that they are numbered in decreasing order along the positive y axis. Similarly, the lower six fringes along the y axis on the right half of the picture in Fig. E6.4b are numbered in increasing order in the positive y axis since $p_m < p_s$.

Because only relative order is important, we assign 100 to the top fringe (relative large number is chosen to avoid negative numbers). The top three fringes are then in orders of 100, 99, and 98 (in decreasing order along the positive y axis), as illustrated in Fig. E6.4c. The fringe orders of 97, 96, and 95 (also shown in Fig. E6.4c) can readily be assigned consecutively. The lower six fringes are assigned numbers of 98, 99, 100, 101, 102, and 103, since they are in increasing order along the positive y axis. If we visualize the moiré fringe pattern as an elevation contour map, the moiré fringes are the contour lines on the map. For example, there are two 98-order fringes because there is a valley between them; the fringe numbers may be the same in different regions (e.g., 96, 97, 98, 99, 100). This is an easy and useful way to check whether the numbering is correct.

6.5   INSTRUMENTATION

In many applications, the moiré fringe pattern is observed by the arrangements shown in Fig. 6.11a for a transparent specimen or in Fig. 6.11b for an opaque specimen. In both arrangements the master grating is placed in contact with the specimen. When the specimen deforms, moiré fringes are produced that can be viewed or photographed by the camera. Unfortunately, it is not always possible to place the master grating in contact with the specimen, for example, in high-temperature applications. One of the best and most versatile instrumentations is a rigid accurately aligned setup where the specimen grating is projected and optically superimposed on the master grating, which is not near the specimen. A moiré fringe pattern can then be observed when the speci-

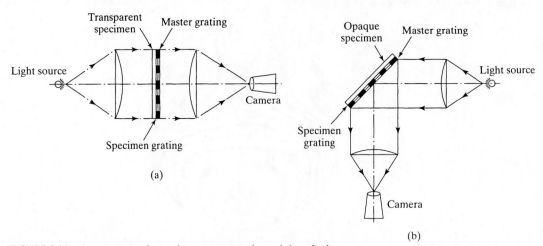

FIGURE 6.11   Arrangements for strain measurement by moiré methods.

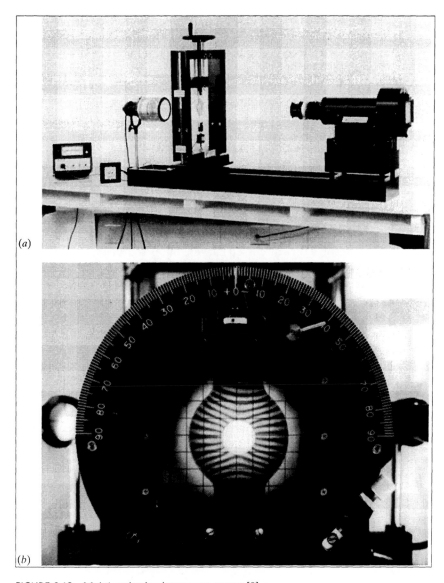

(a)

(b)

FIGURE 6.12   Moiré projection instrument system [8].

men deforms. Since the master grating does not come directly in contact with the specimen, this kind of arrangement can be used in almost any possible test environment. A moiré projection instrument system manufactured by the Measurements Group, Inc. is illustrated in Fig. 6.12a. It can be used for observing and analyzing the moiré fringe patterns at a distance from the specimen. Figure 6.12b shows the moiré fringe pattern of a test specimen observed with the projector.

6.6   OUT-OF-PLANE DISPLACEMENT MEASUREMENTS

In engineering practice, many problems are caused by plates subjected to lateral loadings. The out-of-plane displacement w is thus an important quantity to be measured. In this section the essential features of the out-of-plane displacement measurement by the shadow moiré method are described.

In the shadow moiré method, the specimen grating is not separate (as in the cases of geometric moiré discussed earlier) but is the shadow the master grating casts on the surface of the testing component—thus the name *shadow moiré*. The shadow of the master grating (the specimen grating) will distort as depths of the out-of-plane displacement w increase. Moiré fringes, which are closely related to the topology of the surface, are formed when the shadow is viewed together with the master grating by the naked eye or a camera.

Figure 6.13 shows one of the arrangements used in the shadow moiré method, where a collimated light beam impinges on the master grating (with pitch p_m) at an incident angle of α, and the specimen is viewed at an angle β from the normal to the master grating plane. From Fig. 6.13 it is clear that the segment AB on the master grating occupies the segment AC on the specimen surface; but point C coincides with point D at the image plane of the viewing camera since parallel receiving is used. Therefore, moiré fringes are formed as the result of an interference between the lines contained in AD and the lines contained in AC (or AB).

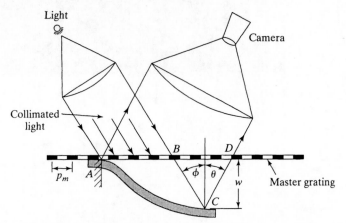

FIGURE 6.13   Arrangement in shadow moiré method [3] (parallel illumination and parallel receiving).

Suppose that there are k lines in AB and m lines in AD; then

$$BD = AD - AB = (m - k)p_m = Np_m$$

where N is the fringe order at point D if the fringe order is zero at point A and p is the pitch. From the geometry illustrated in Fig. 6.13, BD is also given by

$$BD = w(\tan \phi + \tan \theta)$$

where w is the vertical distance from point C to the master grating plane. Thus

$$w = \frac{Np_m}{\tan \phi + \tan \theta} \qquad (6.24)$$

It can be seen from Eq. (6.24) that the moiré fringes are contours of equal vertical distance measured from the master grating plane. In practice, the specimen is viewed from the normal to the master grating plane (i.e., $\beta = 0$). Equation (6.24) becomes

$$w = \frac{Np_m}{\tan \phi} \qquad (6.25)$$

EXAMPLE 6.5

Figure E6.5a shows a moiré fringe pattern obtained by the shadow moiré method. The specimen is a thin, simply supported circular plate with diameter of d, which is subjected to a uniformly distributed loading. Assuming that the pitch is 0.01 in., ϕ is 18.43°, and θ is 0°, determine the deflection curve along a diameter.

Solution   The fringe order at simply supported boundary is zero, since the deflection there is zero. The rest of the fringes can readily be numbered by counting directly in increasing order. The deflection can be obtained by Eqs. (6.24) or (6.25), namely,

$$w = 0.03N \text{ in.}$$

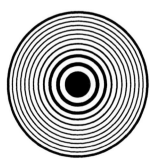

FIGURE E6.5a   [3]

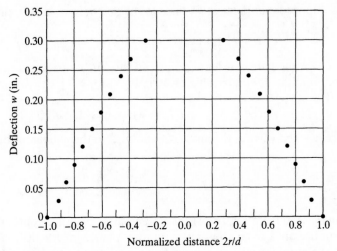

FIGURE E6.5b Deflection curve along a diameter.

The deflection curve along a diameter is shown in Fig. E6.5b, where the normalized distance $2r/d$ is used. It can be seen from Fig. E6.5a that the moiré fringes present a radial symmetry, since the fringes are contours of equal deflection with respect to the master grating plane.

The arrangement with parallel illumination and parallel receiving can only be applied to small models, because it is difficult to produce collimated light with a large field. Therefore, in the case of large models, a shadow moiré method with both point illumination and point receiving is employed. This is shown in Fig. 6.14. Similar to the shadow moiré method, with parallel illumination and parallel receiving, and from the geometry shown in Fig. 6.14, we have

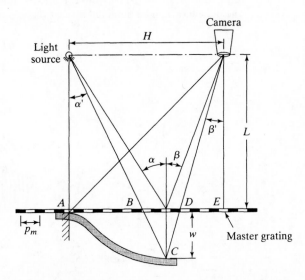

FIGURE 6.14 Arrangement in the shadow moiré method [3] (point illumination and point receiving).

$$BD = AD - AB = (m - k)p_m = Np_m$$

and

$$BD = w(\tan \alpha' + \tan \beta')$$

Thus

$$w = \frac{Np_m}{\tan \alpha' + \tan \beta'} = \frac{Np_m(L + w)}{H} \tag{6.26}$$

or

$$w = \frac{LNp_m}{H - Np_m} \tag{6.27}$$

where p_m, H, and L are known. Once the fringe order N is determined, w can be computed from Eq. (6.27). There are two other possible arrangements: parallel illumination with point receiving and point illumination with parallel receiving. The operating principles are essentially the same and are omitted here.

In an earlier discussion, the master grating plane served as the reference plane for measurement of the out-of-plane displacement induced by load only. In practice, the master grating is placed a short distance away from the specimen. If a point of zero w exists in the specimen and is known beforehand, the master grating can be adjusted to locate a moiré fringe (zero-order fringe) that passes through that point. If no point of zero w exists in the specimen, it may be necessary to determine the out-of-plane displacement w at a convenient point on the specimen by some other means so that the displacement w at all other points on the specimen can be referred to that point.

If the plate surface is not flat or if it is a curved surface before loading, the moiré fringe pattern is formed for an undeformed specimen. To measure the load-induced, out-of-plane displacement, moiré fringe patterns for both undeformed and deformed specimens are recorded. For the arrangement of parallel illumination with parallel receiving, w can be calculated by

$$w = w_2 - w_1 = \frac{N_2 p_m}{\tan \phi} - \frac{N_1 p_m}{\tan \phi} = \frac{(N_2 - N_1)p_m}{\tan \phi} \tag{6.28}$$

where w_1 and w_2 are the initial displacement and deformed displacement measured by the shadow moiré method, w is the load-induced displacement, and N_1 and N_2 are the initial and deformed moiré fringe orders, respectively, at the same point on the specimen.

6.7   OUT-OF-PLANE SLOPE MEASUREMENTS

Determination procedures for out-of-plane displacement by shadow moiré methods were discussed in Section 6.6. Theoretically, the stress distributions can then be calculated by using Eq. (6.29) for a thin plate subjected to a lateral loading or to a bending moment once the displacement is known, namely,

$$\sigma_x = -\frac{Ez}{1 - \mu^2}\left(\frac{\partial^2 w}{\partial x^2} + \mu\frac{\partial^2 w}{\partial y^2}\right) \tag{6.29a}$$

$$\sigma_y = -\frac{Ez}{1 - \mu^2}\left(\frac{\partial^2 w}{\partial y^2} + \mu\frac{\partial^2 w}{\partial x^2}\right) \tag{6.29b}$$

where E is Young's modulus, μ is Poisson's ratio, and z is the distance from the neutral plane of the plate. However, the second derivatives cannot be obtained directly from double differentiations of the measured displacement w with sufficient accuracy. Alternative moiré methods, originated by Ligtenberg [5], have been developed to measure the slopes $\partial w/\partial x$ and $\partial w/\partial y$. A single differentiation of the measured slopes is then used to calculate the stress distributions, and it usually provides reasonably accurate stress values.

The basic method (known as the *indirect reflection moiré method*) for measuring out-of-plane slopes is shown in Fig. 6.15. It is similar to the shadow moiré method except that a master grating is necessary. The specimen grating is actually the virtual image of the master grating, which deforms when the plate is deflected. The plate model must have a mirrored surface in order to reflect the light. A collimated light beam is unnecessary, since the virtual image does not depend on the angle of the incident light. Moiré fringes are formed when the undeformed grating image is superimposed on the deformed grating image by double exposures.

From the geometry of Fig. 6.15, it can be seen that the location of any point d on the master grating viewed by the camera results from the reflection of a typical point A on the specimen surface and shifts to point e when the plate deflects under load. Because points d and e are superimposed on the master grating via point A on the undeformed plate and point A' on the deformed specimen, moiré fringe can be observed. If the number of lines are k and m in segments ad and ae, then

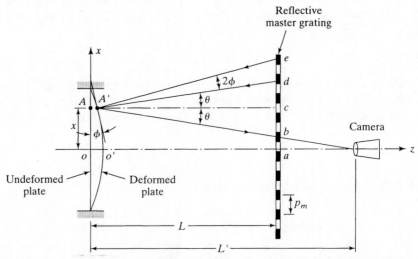

FIGURE 6.15   Optical arrangement of the indirect-reflection moiré method [3].

$$ed = ae - ad = (m - k)p = Np_m$$

Also,

$$ed = ce - cd = L\left[\tan(2\phi + \theta) - \tan\theta\right]$$

Combining the two equations above yields

$$\frac{Np_m}{L} = \frac{\tan 2\phi + \tan\theta}{1 - \tan 2\phi \tan\theta} - \tan\theta \tag{6.30}$$

If ϕ is small, which is usually true for plate bending problems, and noting that $\phi = \partial w/\partial x$, Eq. (6.30) can be simplified as

$$\frac{\partial w}{\partial x} = \frac{Np_m}{2L(1 + \tan^2\theta)} = \frac{Np_m}{2L(1 + x^2/L'^2)} \tag{6.31}$$

If $x << L'$, Eq. (6.31) is reduced to

$$\frac{\partial w}{\partial x} = \frac{Np_m}{2L} \tag{6.32}$$

Since L and p are known, $\partial w/\partial x$ can be obtained by Eq. (6.32) once the fringe order N is determined.

Similarly, to determine the slope $\partial w/\partial y$, we simply turn the master grating 90°. Thus

$$\frac{\partial w}{\partial y} = \frac{N'p_m}{2L} \tag{6.33}$$

To obtain the curvatures, we can follow exactly the same steps as in the displacement method described in Section 6.4 if u and v are replaced by $\partial w/\partial x$ and $\partial w/\partial y$. The bending moments or the stress distributions for a plate can then be determined.

In Eq.(6.31) the term $(x/L')^2$ makes the fringes not exactly the counter lines of slope. To eliminate this term, we either vary the camera position such that $x << L'$ or have a master grating on a cylindrical surface with a radius of $3.5L$ [5]. Figure 6.16 illustrates two arrangements used in practice. In Fig. 6.16a, a hole should be drilled through the grating to allow a view by the camera. In Fig. 6.16b the arrangement is more flexible. The master grating is the image on a large ground glass projected by an ordinary slide projector; therefore, the grating pitch can be changed freely by changing either the projection distance or the grating slide. Also, the light and the grating are provided simultaneously with this arrangement. Figure 6.17 shows moiré fringe patterns for a clamped circular plate subjected to the same uniform pressure but with different grating pitches.

Extensive research has been done to measure the curvatures (thus the bending moments or stresses) directly using moiré methods. The analysis is lengthy and is omitted here. The interested reader is referred to articles by Kao and Chiang [6] for plates and Politch and Gryc [7] for cylindrical shells.

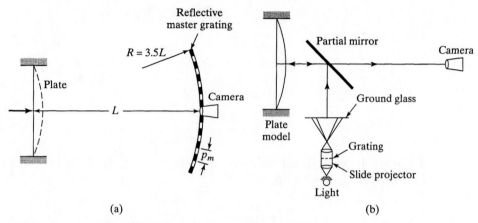

(a)   (b)

FIGURE 6.16   Arrangements for measuring the out-of-plane slope [1,3].

6.8 GRATINGS AND THEIR APPLICATIONS

Obviously, gratings play an important role in strain analysis by moiré methods. To obtain reliable results by these methods, the grating pattern must be of high quality, having uniform pitch, uniform line/space ratio, and straight lines. It must be free of initial distortion when applied to the specimen and have the line density necessary to meet the required sensitivity in the experiment. The line density should always be large enough to measure small strains.

Gratings with a line density of 200 to 1000 lines per inch (lpi), or 8 to 40 lines per millimeter, are commercially available for strain analysis. Measurements Group, Inc. [8] provides type FTG bondable grid (bidirectional lines) sheets, size 4 × 4 in. (100 × 100 mm), with line densities of 200, 500, and 1000 lpi and VPM master grating sheets, size 4 × 4 in. (100 × 100 mm) and 1 × 1 in. (25.4 × 25.4 mm), with line densities of 200, 500, and 1000 lpi. Some suppliers (e.g., optical grating companies [10]) offer gratings with a line density greater than 1000 lpi (or 40 lines per millimeter).

Various techniques are used to produce the specimen gratings. Gratings can be scribed, ruled, etched, stamped, printed, photographed, or bonded to test parts,

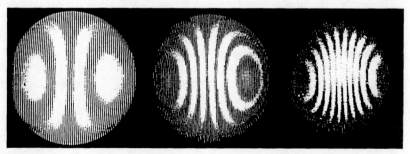

FIGURE 6.17   Fringe patterns for a clamped circular plate under uniform pressure by the reflection moiré method [3].

depending on the application. One successful method is to apply the grating pattern by photographic etching, and it is particularly useful for metallic specimens needed, for example, to measure strains at high temperature.

The *transfer bonding method* is used most widely today for applying the grating pattern to the specimen. This method is accurate, simple, and inexpensive. Furthermore, the transfer bonding method extends the moiré method to the study of materials that cannot be etched and provides the same grating pattern repeatedly [8]. This method, however, cannot be used in high-temperature strain measurements.

Bonding procedures are shown schematically in Fig. 6.18 and explained briefly below.

1. Clean the specimen surface, then apply the adhesive to the bonding area. (Note that thick layer adhesive should be applied if the surface is rough.)
2. Place the grating and its carrier in contact with the adhesive. (It is very important that the grating not be contaminated during cutting and handling.)
3. Cover the grating with a silicon rubber sheet and apply bonding pressure as suggested by the manufacturer with a metal plate during curing of the cement.
4. After the adhesive is completely cured, strip away the carrier, leaving an accurate distortion-free grating pattern on the specimen.

Techniques employed by moiré methods are constantly being improved. For example, the interferometric moiré method [2] increases sensitivity by more than one order of magnitude over the geometric moiré method, and the entire data analysis process can be computerized using the pattern-recognition technique [3,10]. It can be seen why only the very basic principles used in geometric moiré methods are presented here. More details on moiré methods may be found in references 1, 3, 9, 10, and 11. Sciammarella [13] gives an excellent review on the moiré method up to 1981.

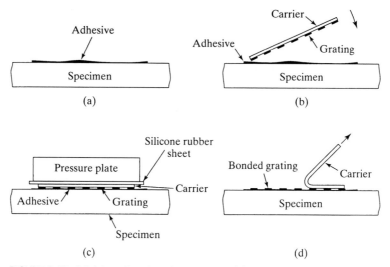

FIGURE 6.18 Moiré gratings bonding procedure [8].

PROBLEMS

6.1. Verify Eq. (6.11).

6.2. Assume that $\delta/p_m = 50$ and the master grating lines are in the x direction, and plot the engineering strain ϵ versus α for $0 \leq \alpha \leq \pi$.

6.3. Derive an expression for deformed grating pitch p_s in terms of α and δ_y, where δ_y is the distance between fringes in the primary direction as shown in Figs. P6.3a and b, when α is an acute angle as well as an obtuse angle. The equation is useful in practice, since it facilitates determining δ_y for some cases.

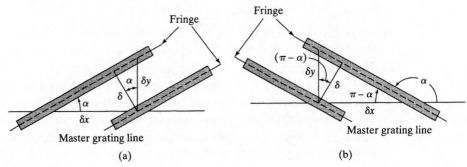

FIGURE P6.3   [1]

6.4. Repeat Problem 6.3 in terms of δ_x instead of δ_y, where δ_x is the distance between fringes in the secondary direction, as also shown in Fig. P6.3.

6.5. A thick epoxy ring was loaded under diametral compression. At a load of 1800 lb, the moiré fringe pattern with the horizontal master grating is as shown in Fig. P6.5. The material constants for the epoxy and the specimen dimensions are also shown in Fig. P6.5.

   (a) Plot the displacement (v)–position (s) curve along the inner boundary. (For simplicity, s is the distance measured on the deformed specimen).

   (b) Verify that $v = 0.2PN/Et$.

6.6. Figure P6.6 shows the moiré fringe pattern for a circular disk under diametral compression. The primary direction is in the y direction. Determine the proper numbers of the fringe orders. Note that the bottom plate was fixed during the experiment.

6.7. Use the moiré fringe pattern shown in Fig. P6.6 to plot the displacement v versus the undeformed position along the y axis (the axis of the symmetry of the disk). The pitch of the grating p_m is 1/300 in. Also determine the distribution of the strain component ϵ_y along the y axis. Note that Fig. P6.6 shows the actual dimension of the deformed disk.

6.8. Determine the distribution of the strain component ϵ_y (the average strain between two fringes) along the x' axis for the moiré fringe pattern shown in Fig. P6.6.

6.9. The shadow moiré method can also be used to determine the mean normal stress distribution in a plane elastic field, which is useful to separate the principal stresses in photoelasticity. The moiré fringes are the loci of points exhibiting the equal transverse strain (isopachics) in a plate. Assume that the length, width, and thickness of the specimen are, respectively, L, w, and h; Young's modulus and Poisson's ratio are E and μ; and the master grating is initially in contact with the specimen and fixed in space. Derive an expression to relate the fringe order with the mean normal stress $(\sigma_1 + \sigma_2)$ for the plane stress specimen.

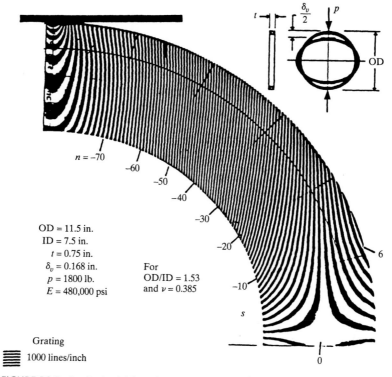

OD = 11.5 in.
ID = 7.5 in.
t = 0.75 in.
δ_v = 0.168 in.
p = 1800 lb.
E = 480,000 psi

For
OD/ID = 1.53
and ν = 0.385

Grating
1000 lines/inch

FIGURE P6.5   Isothetics (v) in a circular epoxy ring [1].

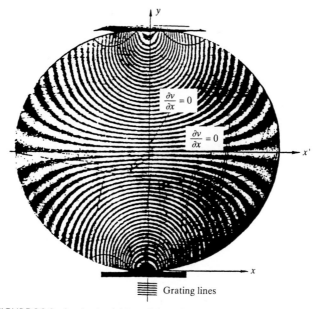

Grating lines

FIGURE P6.6   Isothetics (v) in a disk under diametral compression [1].

6.10. Figure P6.10 shows the shadow moiré fringe pattern for a symmetrically grooved (semi-circular grooves) specimen subjected to uniaxial tension. The specimen was made sufficiently long to secure a uniform stress distribution in the specimen away from the grips.

   (a) Determine the fringe orders. (Note that the dark fringes are half-order fringes.)

   (b) Determine the sum of principal stress distribution (in fringe orders) at the rim of the grooves ($-90° \leq \theta \leq 90°$). For this part, the relationship between N and the sum of principal stress in an elastic plane stress field should be known, as given in Problem 6.9.

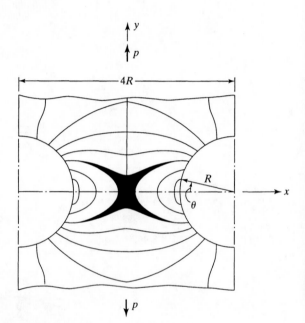

FIGURE P6.10   Isopachics of a symmetrically grooved specimen.

6.11 Determine the sum of the principal stress distribution (in fringe orders) at the minimum section ($0 \leq x/R \leq 1$) of the symmetrically grooved strip subjected to uniaxial tension. The shadow moiré fringe pattern is shown in Fig. P6.10.

REFERENCES

[1] A. J. Durelli and V. J. Parks, *Moiré Analysis of Strain,* Prentice Hall, Upper Saddle River, N.J., 1970.

[2]* D. Post, Moiré interferometry, Chapter 7 in *Handbook on Experimental Mechanics,* A. S. Kobayashi, ed., Prentice Hall, Upper Saddle River, N.J., 1987.

[3] F. P. Chiang, Moiré methods of strain analysis, Chapter 6 in *Manual of Engineering Stress Analysis,* A. S. Kobayashi ed., Prentice Hall, Upper Saddle River, N.J., 1982.

[4] D. Post, The moiré grid-analyzer method for strain analysis, *Exp. Mech.,* **5,** 368–377, 1965.

[5] F. K. Ligtenberg, The moiré method: a new experimental method for determination of moments in small slab models, *Proc. SESA,* **12,** 83–98, 1955.

[6] T. Y. Kao and F. P. Chiang, Family of grating techniques of slope and curvature measurements for static and dynamic flexure of plate, *Opt. Eng., 21,* 721–742, 1982.

[7] J. Politch and S. Gryc, Second derivatives of displacement on cylindrical shells by projection moiré, *Appl. Opt., 28,* 111–118, 1989.

[8] *Moiré Method of Stress Analysis,* Tech Note TN-703, Measurements Group, Inc., Raleigh (North Carolina), 1980.

[9] P. S. Theocaris, *Moiré Fringes in Strain Analysis,* Pergamon Press, Elmsford, N.Y., 1969.

[10]* V. J. Parks, Geometric moiré, Chapter 6 in *Handbook on Experimental Mechanics,* A. S. Kobayashi, ed., Prentice Hall, Upper Saddle River, N.J., 1987.

[11] J. W. Dally and W. F. Riley, *Experimental Stress Analysis,* 2nd ed., McGraw-Hill, New York, 1978.

[12] S. Morse, A. J. Durelli, and C. A. Sciammarella, Geometry of moiré fringes in strain analysis, *J. Eng. Mech. Div. ASCE,* **86**(EM4), 105–126, August 1960.

[13] C. A. Sciammarella, The moiré method: a review, *Exp. Mech., 22,* 418–433, 1982.

---

*See also the same chapters by the same authors of the handbook, second revised edition, A. S. Kobayashi, ed., VCH Publishers, Inc., New York, N. Y., 1993.

C H A P T E R   7

Holographic Interferometry

7.1 INTRODUCTION

Holography is a technique to record a wavefront and subsequently to reconstruct it in the absence of the original wavefront. *Holo* is from the Greek word *holos*, the whole, because all the information (both the wave phase and amplitude) is recorded by the holographic technique. Thus observation of the reconstructed wavefront provides exactly the same physical data as that of the original wavefront, so that a three-dimensional image can be observed as if the original object were still there [1].

Holographic interferometry is an important technique in experimental mechanics. The method is an extension of interferometric measurement techniques and can be applied to static and dynamic displacement measurements for an optically rough surface. It provides a means for nondestructive whole-field displacement measurements with very high sensitivity. Holographic interferometry has received considerable attention and developed rapidly in recent years. It is likely to find even more applications in engineering practice in the future.

In this chapter we introduce only the basic concepts or principles of holography. No attempt is made for completeness, since many applications using the technique are still being developed. The interested reader is referred to the book of Jones and Wykes [1] for a more detailed discussion of theory and applications and to references 2 and 3 for more recent developments and applications.

7.2 INTERFERENCE AND DIFFRACTION

To design or use a holographic system effectively, it is important for a potential holographer to have a basic understanding of the nature of light. Therefore, some additional concepts, such as wavefront, coherent length, interference, and diffraction, will be introduced. In Chapter 4, the nature of light was described briefly. Light waves were interpreted as electromagnetic radiations. By recalling Eq. (4.1), the general form of a light wave traveling in the positive z direction with velocity v can be rewritten as

$$L = A_L \cos\left[\frac{2\pi}{\lambda}(z - vt) + \phi\right] = \text{Re}[A_L e^{j[(2\pi z/\lambda)] - \omega t + \phi]}] \tag{7.1}$$

where A_L is the amplitude, λ the wavelength, t the time, ϕ the phase, ω the circular frequency, $j = \sqrt{-1}$, and Re indicates the real part of the complex variable.

For simplicity in presentation, Re will be omitted, as was done in Chapter 4, and attention should focus on the hologram plane, which is conveniently set at $z = 0$. Since only the coherent monochromatic light laser (i.e., all light waves having the same wavelength) is considered in this chapter, $e^{-i\omega t}$ will also be omitted. Thus Eq. (7.1) can be simplified to

$$L = A_L e^{j\phi} \tag{7.2}$$

where L is the complex amplitude of the light wave, or simply the light wave.

Wavefront. The wavefront is a surface in space along which phase ϕ is constant at all points. For example, the wavefront of the light reflected from a point in air is a spherical surface. When this wavefront strikes the eye, one can see the point that radiated it.

Interference. If two light waves, L_1 and L_2, are simultaneously incident on a plane at $z = 0$, the resulting wave L is given by

$$L = L_1 + L_2 = A_{L1} e^{j\phi_1} + A_{L2} e^{j\phi_2} \tag{7.3}$$

By letting L^* be the complex conjugate of L, the intensity of L can be obtained by Eq. (4.8). For simplicity, C is omitted.

$$\begin{aligned} I = LL^* &= (A_{L1} e^{j\phi_1} + A_{L2} e^{j\phi_2})(A_{L1}^{-j\phi_1} + A_{L2} e^{-j\phi_2}) \\ &= A_{L1}^2 + A_{L2}^2 + 2A_{L1}A_{L2} \cos(\phi_1 - \phi_2) \end{aligned} \tag{7.4}$$

From Eq. 7.4 we find that the result of two light waves may cause constructive or destructive interference, depending on their relative phase. For example, if they are in phase (i.e., $\phi_1 = \phi_2$), the interference is constructive since the brightness will be increased [$I = (A_{L1} + A_{L2})^2$]. If they are out of phase (i.e., $\phi_1 = \phi_2 \pm \pi$), the interference is destructive since the brightness will be decreased [$I = (A_{L1} - A_{L2})^2$]. In general, the light intensity is modulated by the term $\cos(\phi_1 - \phi_2)$ at each point. If the relative phase $(\phi_1 - \phi_2)$ remains fairly constant with time, an interference fringe pattern is observed.

Coherent Length. Light that is capable of interference is known as *coherent*. A coherent light wave is divided by a beam splitter into two light waves traveling along two different paths and then meeting again at a certain point. A stable interference fringe pattern is formed if the difference between the two path lengths is within a certain limit. That limit is called the *coherent length*. For an ordinary light source, the coherent length is only about 1 mm, but for a laser it is several feet. That explains why holography did not receive much attention in experimental mechanics until lasers were developed in the 1960s.

Diffraction. When a light wave is incident on an aperture, an edge, or an optical rough surface, the light is spread out in a way not consistent with the law of reflection and refraction, described in Chapter 4. For example, if an opaque plate with a large hole is placed between a light source and a viewing screen, the illuminated area on the screen is due to the rays drawn from the light source through the hole boundary and reaching the screen. As expected, when the hole is made smaller and smaller, the illuminated area on the screen becomes smaller and smaller. However, the illuminated area on the screen reaches its minimum at a certain size of hole and then increases if the size of the hole decreases further. Obviously, there is bending in the light beam. In other words, light waves do not always travel in straight lines. This phenomenon is known as *diffraction*.

Diffraction can be explained by the Huygen–Fresnel principle, derived empirically and later confirmed analytically. The principle states that "each point on a wavefront acts as a point source of so-called secondary waves. The amplitude of the wavefront as it propagates is given by the sum of these secondary waves" [1].

7.3 WAVEFRONT RECORDING AND RECONSTRUCTION BY HOLOGRAPHY

An ordinary photograph records only information regarding the amplitude of light waves, and all phase information is missing. This flat picture of a three-dimensional source is based on the light-intensity amplitude in the original source. Holography, on the other hand, records the entire light-wave field (both amplitude and phase information contained in light waves scattered by an object) on a flat plate due to the light interference. After being exposed and developed properly, the plate becomes a hologram. After reconstructing the wavefront stored on the hologram, a three-dimensional image of the object is obtained. Typical arrangements for making holograms are shown in Fig. 7.1.

The arrangement in Fig. 7.1a is used when the object is opaque; the arrangement in Fig. 7.1b is used for a transparent test sample. The laser beam is divided into two light beams by a beam splitter. One beam, the *reference beam*, is reflected by a mirror and reaches the hologram screen directly; the other, the *object beam*, is also reflected by a mirror and then illuminates the object. The light is either reflected from (Fig. 7.1a), or transmitted through (Fig. 7.1b), the test sample and finally reaches the hologram screen. Two light beams interact at H, a high-resolution photosensitive plate, and give an interference fringe pattern. The plate is then exposed and properly developed into a hologram that records all light-beam data scattered by the original test object [i.e., both the amplitude (the light intensity distribution) and phase information].

Contrary to an ordinary photograph, the image of the test object cannot be seen directly from a hologram. However, it is easy to reconstruct the wavefront by illuminating the hologram with only the reference beam, as shown in Fig. 7.2. Since a hologram is a diffraction pattern, the reference light will be diffracted by the hologram to form wavefronts that are identical to those originally resulting from the test object itself. If the reconstructed wavefronts strike our eyes, we have the illusion that the test object is physically present there.

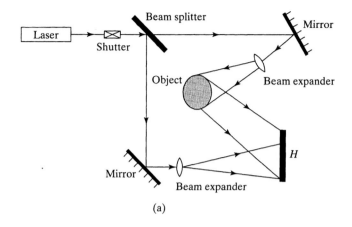

(a)

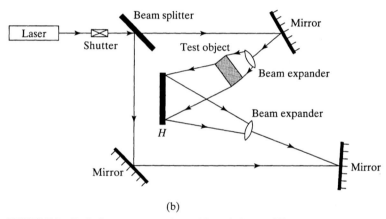

(b)

FIGURE 7.1   Typical arrangements for making a hologram [3].

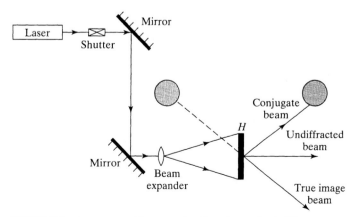

FIGURE 7.2   Arrangement for reconstruction of wavefronts [3].

The wavefront recording and reconstruction process by holographic interferometry can be described mathematically as follows. Let the object and reference beams be

$$L_o = A_{Lo}e^{j\phi_o} \qquad L_r = A_{Lr}e^{j\phi_r}$$

where subscripts o and r represent the object and reference light beams, respectively. When these two beams are incident on the photosensitive plate H simultaneously, the resulting wave L is computed by Eq. (7.3) and given by

$$L = L_o + L_r = A_{Lo}e^{j\phi_o} + A_{Lr}e^{j\phi_r}$$

The light intensity of L is determined by Eq. (7.4):

$$\begin{aligned} I = LL^* &= (A_{Lo}e^{j\phi_o} + A_{Lr}e^{j\phi_r})(A_{Lo}e^{-j\phi_o} + A_{Lr}e^{-j\phi_r}) \\ &= A_{Lo}^2 + A_{Lr}^2 + A_{Lo}A_{Lr}[e^{j(\phi_o-\phi_r)} + e^{-j(\phi_o-\phi_r)}] \\ &= A_{Lo}^2 + A_{Lr}^2 + 2A_{Lo}A_{Lr}\cos(\phi_o - \phi_r) \end{aligned}$$

Therefore, the photosensitive plate will be exposed with a light intensity distribution defined by Eq. (7.5). It is clear that the light intensity contains both the amplitude A_{Lo} and phase ϕ_o of the object beam. At each point on the plate, the light intensity is modulated by the cosine of the difference in phase between the object and the reference beams.

As explained previously, after being properly developed, a hologram is obtained that is a diffraction pattern source. During the reconstruction process, the reference beam is incident on the hologram, and a part of the light will be transmitted and a part of the light will be absorbed. The transmittance T is defined by

$$T = \frac{\text{Amplitude of the transmitted light}}{\text{Amplitude of the incident light}} \tag{7.6}$$

which is assumed to be a linear function of light intensity I. For simplicity of presentation, the constant of proportionality is set as unity, since it will only affect the brightness and not the fringe pattern in the analysis. Accordingly, the amount of reference light that transmits through the hologram, L_T, is given by

$$L_T = L_r T = L_r I \tag{7.7}$$

Substituting Eq. (7.5) into Eq. (7.7) yields

$$L_T = A_{Lr}(A_{Lo}^2 + A_{Lr}^2)e^{j\phi_r} + A_{Lo}A_{Lr}^2 e^{j\phi_o} + A_{Lo}A_{Lr}^2 e^{-j(\phi_o-2\phi_r)} \tag{7.8}$$

Equation (7.8) indicates that the reference beam L_r has been regenerated by diffraction from the hologram. The first term represents the undiffracted beam (the zero-order diffraction wave), which travels along the original direction of the reference beam. The second term represents the true-image beam (the first-order diffraction wave), which contains both amplitude A_{Lo} and phase information ϕ_o of the object beam and travels along the original direction of the object beam. The last term represents the conjugate-image beam (another first-order diffraction wave), which also contains both amplitude A_{Lo} and phase information $-\phi_o$ of the object beam. All three beams are shown schematically in Fig. 7.2.

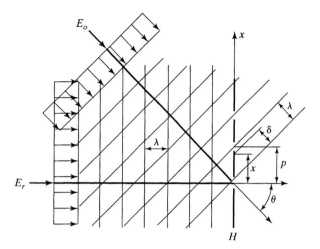

FIGURE 7.3   Interference of two plane waves [3].

To explain further the process of wavefront recording and reconstruction in holographic interferometry, we assume that both object and reference beams are plane waves of equal constant amplitude ($A_{Lo} = A_{Lr} = A_L$) which interact at H, as shown in Fig. 7.3. Since only the difference in phase between the reference and object beams is important, we assume that $\phi_r = 0$. Thus the phase difference is ϕ_o, a function of x, as shown in Fig. 7.3. The phase difference, ϕ_o, can be determined as

$$\phi_o(x) = \frac{2\pi\delta}{\lambda} = \frac{2\pi \sin\theta}{\lambda}x \qquad (7.9)$$

where δ, λ, and θ are the linear phase, wavelength, and angle between the two normals of plane waves, respectively.

Using Eq. (7.5), the distribution of light intensity on plate H is now given by

$$I = 2A_L^2 + 2A_L^2 \cos\left(\frac{2\pi \sin\theta}{\lambda}x\right) \qquad (7.10)$$

It can be shown that maximum intensity will occur whenever the difference in phase between the object and reference beams is $2n\pi$ ($n = 0,1,2, ...$) and that minimum intensity will occur whenever the difference in phase between the object and reference beams is $(2n + 1)\pi$ ($n = 0,1,2, ...$). The distance between consecutive maxima (or minima) is known as the pitch p; thus

$$p = \frac{\lambda}{\sin\theta} \qquad (7.11)$$

In this simple case, the interference fringes are straight lines equally spaced at a distance p. After being exposed and developed properly, a hologram is made that is actually a diffraction grating with pitch p. When a coherent-light laser is incident on the plate, diffraction will occur. It should be pointed out that Eq. (7.11) can also be used to calculate the resolution of the hologram for a given arrangement, as shown in Fig. 7.1a, or to determine the angle θ for a required resolution.

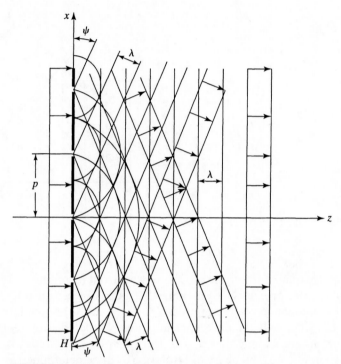

FIGURE 7.4   Diffraction of plane waves by a hologram [3].

Consider now that the hologram is illuminated by the reference beam (plane wave with $\phi_r = 0$ is assumed), as shown in Fig. 7.4. Diffraction occurs which can be explained by the Huygen–Fresnel principle. The semicircles represent the secondary waves, which are rather weak in general and reinforce each other to form two appreciable waves at angles ψ and $(\pi - \psi)$. As is known, if the linear phase difference between two waves is a wavelength λ, reinforcement will occur. From Fig. 7.4 we have

$$p = \frac{\lambda}{\sin \psi} \tag{7.12}$$

Comparing Eq. (7.12) with Eq. (7.11) yields

$$\psi = \theta \tag{7.13}$$

Equation (7.13) shows that the direction of the first-order diffraction wave is coincident with that of the object beam. By using the arrangement shown in Fig. 7.2, we can reconstruct the object wavefronts so that we can "see" the test object as if it were still there.

Furthermore, since $A_{Lr} = A_{Lo} = A$ and $\phi_r = 0$, using Eq. (7.8) directly, the transmitted light is given by

$$L_T = 2A^3 + A^3 e^{j\phi_o} + A^3 e^{-j\phi_o} \tag{7.14}$$

where the first term represents the undiffracted beam, the second term represents the true image beam (first-order diffraction), and the third term represents the conjugate beam, as shown in Fig. 7.4.

EXAMPLE 7.1

The wavelength for a helium–neon laser is 632.8 nm. If the angle θ shown in Fig. 7.3 is 20°, what is the pitch p?

Solution   By Eq. (7.11), the pitch can be computed as

$$p = \frac{\lambda}{\sin\theta} = \frac{632.8 \text{ nm}}{\sin 20°} = 1.85 \ \mu\text{m}$$

Thus the pitch p is 1.85 μm or equivalent to approximately 540 lines/mm.

7.4 DISPLACEMENT MEASUREMENT BY HOLOGRAPHIC INTERFEROMETRY

One of the most important applications of holographic interferometry to experimental stress analysis is the measurement of small in-plane and out-of-plane surface displacements in a test object, subjected to mechanical or thermal loadings. In addition to the advantages of the full-field nature of the measurement and the high sensitivity, the method does not require any special surface preparation, so that it can be applied to any structural surface, even to nonoptical surfaces, such as those of composite materials. In displacement measurements by holographic interferometry, the commonly used techniques are double-exposure holographic interferometry and real-time holographic interferometry.

Double-Exposure Holographic Interferometry

Double-exposure holographic interferometry is a convenient and effective method to measure the surface displacement, especially the out-of-plane displacements. As observed experimentally, two or more holograms can be stored in one photosensitive plate with little apparent deterioration in quality [3]. By using the arrangements shown in Fig. 7.1, double exposures are made, one to record the wavefronts associated with the undeformed test object and the other to record those wavefronts associated with the deformed test object. The two holograms are stored on the same photosensitive plate, and the plate is developed properly. Upon reconstruction, the two images will interfere with each other due to a change in optical length of the object beam and will produce a set of fringes representing contours of displacement in the view direction.

Let the object beam before and after deformation be L_{ob} and L_{oa},

$$L_{ob} = A_{Lo}e^{j\phi_b} \qquad L_{oa} = A_{Lo}e^{j\phi_a} \tag{7.15}$$

where ϕ_b and ϕ_a are the phase of object beam before and after deformation, respectively. Let the reference beam be L_r,

$$L_r = A_{Lr}e^{j\phi_r} \tag{7.16}$$

By using Eq. (7.3), the resultant waves incident on the plate before and after deformation are

$$L_b = L_{ob} + L_r = A_{Lo}e^{j\phi_b} + A_{Lr}e^{j\phi_r}$$
$$L_a = L_{oa} + L_r = A_{Lo}e^{j\phi_a} + A_{Lr}e^{j\phi_r}$$

(7.17)

Assume that the time for both exposures is exactly the same and set to unity for simplicity. The recorded light intensity on the hologram is then given by

$$
\begin{aligned}
I &= L_b L_b^* + L_a L_a^* \\
&= 2(A_{Lo}^2 + A_{Lr}^2) + A_{Lo}A_{Lr}e^{j(\phi_b - \phi_r)}\left[1 + e^{j(\phi_a - \phi_b)}\right] \\
&\quad + A_{Lo}A_{Lr}e^{-j(\phi_b - \phi_r)}\left[1 + e^{-j(\phi_a - \phi_b)}\right]
\end{aligned}
$$

(7.18)

After being developed properly, the hologram is illuminated by the reference beam L_r, which is the same as the one used in the formation process. It should be pointed out, however, that in the reconstruction process, the illuminating light is not necessarily the same as the reference beam used to make the hologram. Any coherent light can be used in the reconstruction process. Since the transmittance is assumed to be a linear function of intensity I, substituting Eqs. (7.16) and (7.18) into Eq. (7.7) yields

$$
\begin{aligned}
L_T &= 2A_{Lr}(A_{Lo}^2 + A_{Lr}^2)\,e^{j\phi_r} + A_{Lo}A_{Lr}^2 e^{j\phi_b}\left[1 + e^{j(\phi_a - \phi_b)}\right] \\
&\quad + A_{Lo}A_{Lr}^2 e^{-j(\phi_b - \phi_r)}\left[1 + e^{-j(\phi_a - \phi_b)}\right] \\
&= L_1 + L_2 + L_3
\end{aligned}
$$

(7.19)

where L_1 is the undiffracted beam, L_2 the true image beam, and L_3 the conjugate image beam, respectively.

The intensity of the true image beam is given by

$$
\begin{aligned}
I &= L_2 L_2^* \\
&= 2A_{Lo}^2 A_{Lr}^4\left[1 + \cos(\phi_a - \phi_b)\right] \\
&= 2A_{Lo}^2 A_{Lr}^4\,(1 + \cos\Delta) \\
&= A\cos^2\frac{\Delta}{2}
\end{aligned}
$$

(7.20)

where A is a constant associated with the amplitude and Δ is the phase difference. From Eq. (7.20) it can be seen that at each point, the intensity of the true image by double-exposure holographic interferometry is modulated by the cosine of the phase difference of the two object beams before and after deformation. When $\Delta = 2N\pi$ ($N = 0, 1, 2, ...$), I is maximum, so that light fringes are observed; when $\Delta = (2N + 1)\pi$ ($N = 0, 1, 2, ...$), I is zero, so that dark fringes are observed. Therefore, an interference fringe pattern is obtained and Δ can be determined by knowing the fringe order N.

The object deformation causes a change in the optical length, which is the distance from the laser to the object and then to the hologram. The change in the optical length, called the *linear phase difference* and denoted by δ, is related to the phase difference Δ by the equation

$$\Delta = \phi_a - \phi_b = \frac{2\pi}{\lambda}\delta$$

where λ is the wavelength of the laser.

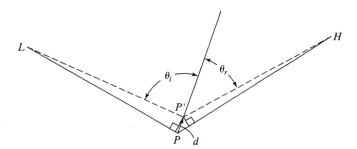

FIGURE 7.5   Optical length change diagram at point P.

To find the relationship between linear phase difference and displacement, consider a point P on an opaque object surface. After deformation, it moves to point P', as shown in Fig. 7.5. In the figure, L is the light source, H the point in the hologram plane, d the displacement at point P (assumed very small), and θ_i and θ_r the incident and reflective angles at point P, respectively. Before deformation, the optical length is $LP + PH$. After deformation, the length is $LP' + P'H$. Since d is very small compared with the optical length, the change in θ_i and θ_r can be neglected. Therefore,

$$\delta = (LP' + P'H) - (LP + PH)$$
$$= d(\cos \theta_i + \cos \theta_r) \tag{7.21}$$

Thus,

$$\Delta = \frac{2\pi d}{\lambda}(\cos \theta_i + \cos \theta_r) \tag{7.22}$$

In vector notations, Eq. (7.22) can be rewritten as

$$\Delta = \frac{2\pi}{\lambda}(\mathbf{pl} + \mathbf{ph}) \cdot \mathbf{PP'} \tag{7.23}$$

where $\mathbf{pl}$ and $\mathbf{ph}$ are unit vectors along the directions of PL and PH, and $\mathbf{PP'}$ is the displacement vector at point P. Equation (7.23) is a general form involving three displacement components for a three-dimensional problem. In principle, three holograms (only when zero-order fringes can be observed on the object surface) or one hologram seen in three different directions is sufficient to determine the displacement components.

One of the special applications of double-exposure holographic interferometry is the out-of-plane displacement measurement. In this case the illumination and observation directions are approximately coincident with the normal of the plate. A practical arrangement is shown in Fig. 7.6. An example of the interference fringe pattern, obtained by double-exposure holographic interferometry, is shown in Fig. 7.7 for a clamped aluminum circular plate subjected to a lateral uniform pressure. The fringes represent the counters of equal out-of-plane displacement.

Since $\theta_i = \theta_r = 0$,

$$d = \frac{\lambda \Delta}{4\pi} = \frac{2N + 1}{4}\lambda \qquad N = 0,1,2, \ldots \tag{7.24}$$

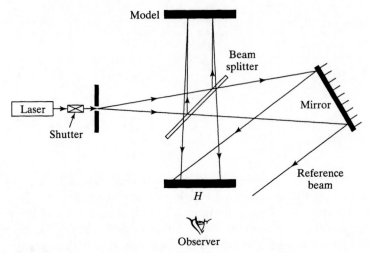

FIGURE 7.6 Arrangement for out-of-plane displacement measurement by double-exposure holographic interferometry [4].

where N is the dark fringe order. When the wavelength λ is known for a particular laser [e.g., for a helium–neon laser $\lambda = 6328$ Å (or 24.91 μin.)], d can be calculated by Eq. (7.24) if the fringe order N is known.

EXAMPLE 7.2

Figure 7.7 shows the double-exposure holographic fringe pattern for a clamped circular plate subjected to lateral pressure. Determine the deflection at the center of the plate. Assume that a helium–neon laser was used in the experiment and the illumination and observation directions are normal to the plate.

FIGURE 7.7 Holographic interference fringes for a circular plate by the double-exposure method [5].

Solution The zero-order fringe should be at the edge of the plate, since deflection is zero there. Also, the deflection is increasing from the edge to its center along the radial direction. By counting directly, we find that the fringe order at the center of the plate is approximately 17.5. Since a helium–neon laser was used, the wavelength is 24.91 μin., and since the directions of illumination and observation are normal to the plate, using Eq. (7.24), we have

$$d = \frac{(2)(17.5 + 1)}{4} (24.91)$$

$$= 224.2 \ \mu\text{in.}$$

However, it is not easy to determine the fringe order in very complicated cases since even zero-order fringes may not exist. This difficulty can be solved with the help of the real-time holographic interferometry, which is discussed next.

Real-Time Holographic Interferometry

By using the arrangements shown in Fig. 7.1, an exposure is made on a photosensitive plate for the undeformed model. After proper developing, the plate is then replaced exactly in its original position. The plate is now illuminated by both the object and reference beams, causing the brightness of the image to increase. When the test model is slowly loaded, it will be deformed. A change will occur in the optical length of the object beam, so that an interference fringe pattern by the two object beams can be observed. This can be recorded by ordinary photography if needed. Since the interference fringe pattern changes with changes in the load applied, increasing or decreasing directions of fringe orders can easily be recognized. Subsequently, the fringe orders at each loading stage can easily be recognized. This technique is known as *real-time holographic interferometry*.

Let the undeformed object beam, the deformed object beam, and the reference beam be L_o, L_d, and L_r, respectively:

$$L_o = A_{Lo}e^{j\phi_o} \qquad L_d = A_{Lo}e^{j\phi_d} \qquad L_r = A_{Lr}e^{j\phi_r}$$

Using Eq. (7.4), the recorded intensity for the undeformed model on the plate is given by

$$I = A_{Lo}^2 + A_{Lr}^2 + A_{Lo}A_{Lr}\left[e^{j(\phi_o - \phi_r)} + e^{-j(\phi_o - \phi_r)} \right] \qquad (7.25)$$

The plate is placed back exactly in its original position. It is illuminated by both the object and reference beams. When the model is deformed, the object beam becomes L_d. By Eq. (7.7) the light transmitted is given by

$$\begin{aligned} L_T &= (L_d + L_r)I \\ &= L_1 + L_2 + L_3 \end{aligned} \qquad (7.26)$$

where L_2 is related to the two object beams and is given by

$$L_2 = A_{Lo}\left[(A_{Lo}^2 + A_{Lr}^2)e^{j\phi_d} + A_{Lr}^2 e^{j\phi_o} \right] \qquad (7.27)$$

The intensity of L_2 may be written as

$$I = L_2 L_2^*$$
$$= A_{Lo}^2 [(A_{Lo}^2 + A_{Lr}^2)^2 + A_{Lr}^4 + 2A_{Lr}^2 (A_{Lo}^2 + A_{Lr}^2) \cos (\phi_d - \phi_o)] \qquad (7.28)$$

When $A_{Lr}^2 \gg A_{Lo}^2$, Eq. (7.28) reduces to

$$I = 2A_{Lo}^2 A_{Lr}^4 (1 + \cos \Delta)$$
$$= A \cos^2 \frac{\Delta}{2} \qquad (7.29)$$

which is exactly the same as Eq. (7.20).

Real-time holographic interferometry is conceptually simple. However, due to the experimental difficulty of replacing the developed plate exactly in its original place, it is commonly used only for qualitative observations, such as determining the zero-order fringe and the increasing or decreasing directions of fringe orders in a complicated case.

7.5   VIBRATION ANALYSIS BY TIME-AVERAGE HOLOGRAPHY

Time-average holography is ideally suited to steady-state vibration problems. The technique provides information for the primary parameters in vibration analysis, such as amplitude distribution, and vibration models. Experimental measurements by time-average holography are noncontact to the test sample; therefore, effects caused by the additional mass in some conventional vibration measurements can be eliminated completely.

In time-average holography, a hologram is made using the arrangement shown in Fig. 7.1 for general cases or that in Fig. 7.6 for normal incident cases to record the vibrating surface of the object. The key feature is that the exposure time should be several times greater than the period of vibration [6]. Conceptually, it is equivalent to many-exposure holographic interferometry to record the object wavefronts at every position occupied by the object during the exposure time on the same hologram. Thus when those wavefronts stored on the hologram are reconstructed, they form an interference fringe pattern.

Consider a point P on a sample, which is vibrating harmonically according to the term $A \cos \omega t$. By Eq. (7.22), the phase change of the light scattered from this point is given by

$$\Delta = \frac{2\pi A \cos \omega t}{\lambda} (\cos \theta_i + \cos \theta_r) \qquad (7.30)$$

where A is the amplitude at point P and θ_i and θ_r are angles of incident and reflected lights with the direction of vibration, respectively. Let the exposure time be t_e and the object beam and the reference beam at the hologram plane be L_{ot} and L_r.

$$L_{ot} = A_{Lo} e^{j(\phi_o + \Delta)} = L_o e^{j\Delta}$$
$$L_r = A_{Lr} e^{j\phi_r}$$

where L_o is the object beam when the sample is stationary and Δ is given by Eq. (7.30). The light intensity in the hologram plane at time t is given by

$$
\begin{aligned}
I(t) &= L_r L_r^* + L_{ot} L_{ot}^* + L_{ot} L_r^* + L_r L_{ot}^* \\
&= (A_{Lo}^2 + A_{Lr}^2) + L_r^* L_o e^{j\Delta} + L_r L_o^* e^{-j\Delta}
\end{aligned}
\tag{7.31}
$$

The hologram effectively records a continuous distribution of frozen fringe patterns corresponding to the model in its vibration cycle, since the exposure time t_e is much greater than the period of vibration τ ($= 2\pi/\omega$). The actual light intensity in the hologram is the average of the intensity $I(t)$ over the period τ. In Eq. (7.31), the second term, $L_r^* L_o e^{j\Delta}$, is of primary importance since it is associated with the true image beam. When only the average of this term over a period is to be considered,

$$
\begin{aligned}
I_{av} &= \frac{L_r^* L_o}{\tau} \int_0^\tau e^{j\Delta}\, dt \\
&= \frac{L_r^* L_o}{2\pi} \int_0^{2\pi} e^{j(2\pi A/\lambda)(\cos\theta_i + \cos\theta_r)\cos\omega t}\, d(\omega t) \\
&= L_r^* L_o J_0\left[\frac{2\pi A}{\lambda}(\cos\theta_i + \cos\theta_r)\right]
\end{aligned}
\tag{7.32}
$$

where $J_0(x)$ is the Bessel function of order zero of the first kind.

After being developed properly, the hologram is illuminated by the reference beam. Using Eq. (7.7), the true image beam is given by

$$
L_{tr} = L_r I_{av} = L_r^2 L_o J_0\left[\frac{2\pi A}{\lambda}(\cos\theta_i + \cos\theta_r)\right]
\tag{7.33}
$$

The intensity of the wavefronts reconstructed will be

$$
I_{tr} = L_{tr} L_{tr}^* = I_o J_0^2\left[\frac{2\pi A}{\lambda}(\cos\theta_i + \cos\theta_r)\right]
\tag{7.34}
$$

where I_o is a constant and is the intensity of the reconstructed wavefront when the object is stationary. If the arrangement shown in Fig. 7.6 is used, $\theta_i = \theta_r = 0$. Equation (7.34) is reduced to

$$
I_{tr} = I_o J_0^2 \frac{4\pi A}{\lambda}
\tag{7.35}
$$

Equation (7.34) or Eq. (7.35) indicates that the intensity of the reconstructed object image is modulated by $J_0^2(x)$. Figure 7.8 shows the variation in the first several cycles of $J_0^2(x)$ with x. It is clear that the brightest region of the reconstructed object image is at $x = 0$, which corresponds to the nodal region (amplitude $A = 0$) in the vibrating pattern. Subsequent maxima ($N = 1,2,3, \dots$) correspond to light fringes and define the contours of constant amplitude, while every root of $J_0(x)$ corresponds to a dark fringe. The amplitude A at the position of a dark fringe is given by

$$
A = \frac{\lambda x_N}{2\pi(\cos\theta_i + \cos\theta_r)}
$$

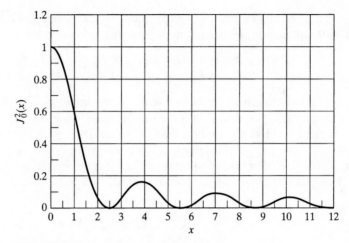

FIGURE 7.8   Distribution of $J_0^2(x)$ with x.

or

$$A = \frac{\lambda x_N}{4\pi}$$

where x_N (N = 0,1,2, ...) are the roots of $J_0(x) = 0$ and N is the dark fringe order. The first five positive nonzero values of x_N are 2.4048, 5.5021, 8.6537, 11.7915, and 14.9309, respectively.

An example of using the time-average technique to identify the vibration mode is shown in Fig. 7.9 [1]. A cantilever metal strip is vibrating in the mode shown in Fig.

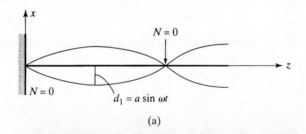

(a)

FIGURE. 7.9   Vibrating mode shape for a cantilever strip [1].

(b)

7.9a. The corresponding fringe pattern in the x–y plane, using the time-average technique, is shown in Fig. 7.9b. The fringes define contours of constant out-of-plane vibrating amplitude. The regions of maximum intensity are the nodal regions (i.e, $N = 0$).

Time-average holographic interferometry has some disadvantages. The phase information is lost due to the averaging of displacement over a cycle; it cannot be used for scanning the full range of frequencies, especially in real time, which is useful for detecting high-order resonant modes, and light intensity decreases rapidly with increasing amplitude so that the higher-order fringes cannot be seen clearly. To overcome such shortcomings in practice, the time-average technique is often combined with double-exposure or real-time holographic techniques.

Consider, for example, the case when time average is combined with the double-exposure technique. A hologram is first made when the object is stationary and the corresponding exposure time is t_1. A second exposure is taken on the same hologram using the time-average technique for the vibrating object, with an exposure time of t_2. The intensity of the reconstructed object image for the normal viewing position is given by

$$I_{tr} = I_o \left(J_0 \frac{4\pi A}{\lambda} + \frac{t_1}{t_2} \right)^2 \tag{7.36}$$

By controlling the ratio of t_1/t_2, usually less than unity in practice, the fringe number is reduced and the brightness of the light fringes is increased so that the fringe pattern can be clearly seen. If $t_1/t_2 = 0.5$, for example, the fringe number is half that obtained by ordinary time-average holography, and the light fringes are brighter.

Other holographic interferometry methods are used in vibration analysis. Two of the most common are stroboscopic or real-time [1,7] and dual-pulsed [1] holographic interferometry. The interested reader will find details in the book by Jones and Wykes [1] and the article by Hazell and Liem [7].

7.6 BASIC EQUIPMENT AND TECHNIQUE FOR MAKING A HOLOGRAM

The basic equipment for holography includes a laser, an isolation bench, recording medium to make a hologram, and some other optical element (beam splitter, mirror, etc.). The most important item is the laser, which provides the necessary coherent light. The laser for holographic applications should be powerful enough that the exposure time is very short. For example, the helium–neon laser with a wavelength of 6328 Å has an output of 5 to 20 mW. When it is operating in the TEM_{∞} (i.e., the laser concentrates its output in the fundamental mode [3]), its coherence length is approximately equal to the length of the laser tube.

It is necessary to isolate the entire apparatus completely from room vibrations during exposure, since any small vibration would change the optical path lengths. If such a change, even a fraction of the wavelength of the light, were to occur, the interference fringes would move during exposure, causing a blurring of the recording hologram. Therefore, an isolation bench is also very important in holography to protect the arrangement from any vibration in the room. In practice, there is no single unique way to construct an isolation bench. One example is a heavy mass of steel supported on a soft-spring system [3].

The recording medium necessary to make a hologram must be a high resolution plate so that the extremely fine fringes (the pitch is in the same order as the wavelength of light) can be resolved. The Agfa-Gevaert 10E75 plate and Kodak 649F spectroscopic plate are two that are commonly used [3]. Other necessary optical elements include partial mirrors and fully reflecting surface mirrors (as shown in Fig. 7.1), pinhole filters, and ordinary microscope objective lenses. They do not need to be of the highest quality.

Before making a hologram, it is recommended that the stability of the isolation system be tested. The laser and mirrors are so arranged that a Michelson interferometer, shown in Fig. 7.10, is formed. The laser beam is expanded and collimated by an ordinary microscope object lens; the light is divided into two beams by the beam splitter B and are reflected by the surface mirrors S_1 and S_2. The two reflected beams are recombined on a ground glass G so that interference fringes can be observed. By observing the stability of the interference fringe pattern in the Michelson interferometer, one can judge the adequacy of the isolation system. If the variation of fringe orders, caused by instability, is less than one-fourth of a fringe at a typical location, good holograms may be obtained.

7.7 COMPARATIVE HOLOGRAPHIC INTERFEROMETRY

Comparative holographic interferometry is a rather new technique that broadens the scope of holographic interferometry in nondestructive inspection [8–10]. The method itself is conceptually simple. Two specimens are used: the master specimen, which is flaw-free; and the test specimen. By comparing the mechanical responses (e.g., displacement or deflection) of two nominally identical specimens subjected to the same loadings, we can indirectly infer the presence of any flaws in the test sample, since the mechanical behavior exhibited by all identical specimens should be the same.

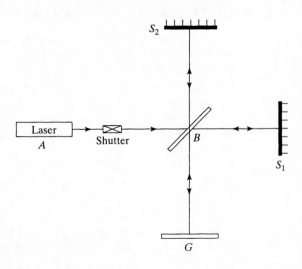

FIGURE 7.10   Michelson interferometer [3].

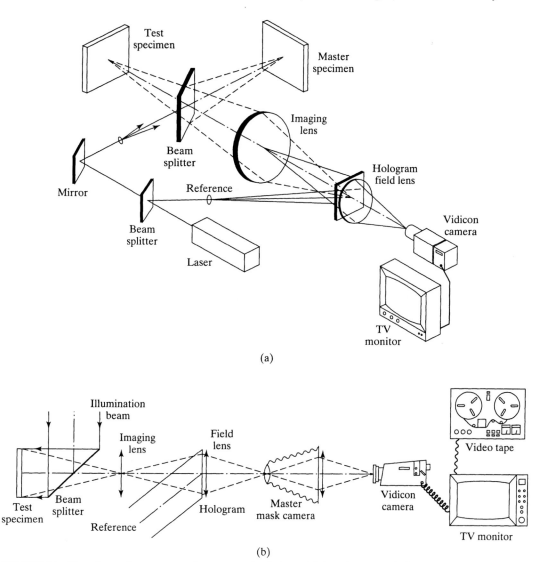

FIGURE 7.11 Experimental arrangement for comparative holographic interferometry [10].

The experimental arrangements of comparative holographic interferometry can be classified into two types [10]. One requires full-time access to the master specimen, as shown in Fig. 7.11a; the other stores the master's mechanical response under a certain loading for a particular holographic setup or "fringeprint" in the system's memory, as shown in Fig. 7.11b. In Fig. 7.11a, the object beam from the master specimen and the object beam from the test specimen are superimposed on the image plane by means of an appropriate optical device. Moiré fringes are formed, which are governed [10] by the equation

FIGURE 7.12 Moiré fringe pattern for a plate with a flaw [10].

$$\Delta w = \frac{N_m \lambda}{2}$$

where Δw is the difference in deflection, N_m the moiré fringe number, and λ the wavelength of the light. We can also visualize evolution of the fringes with increasing loads in real time on a television monitor. An example of this moiré fringe pattern for a test sample with a flaw is shown in Fig. 7.12. In the figure, a closed set of moiré fringes clearly indicates the presence and location of the flaw (indicated by a cross) in the test sample. Note that auxiliary displacement fringes were added to obtain a good formation of the fringe patterns. This is an example of applying the comparative holographic interferometry to nondestructive inspection.

One of the disadvantages of using this type of setup is that the master specimen is subjected to repeated loadings, which can alter the mechanical behavior of the master specimen. It may therefore be suitable for a particular problem but not suitable for routine testing. In Fig. 7.11b, the mechanical response (e.g., displacement) of the master specimen under a predefined load is stored by means of an encoded interference mask. The disadvantage of exposing the master specimen to repeated loadings is thereby eliminated, so it can be used for routine testing. In addition to its application in flaw detection, comparative holographic interferometry may also be used to evaluate material fatigue. The interested reader is referred to the article by Rastogi [10] for more details.

PROBLEMS

7.1. Using Eq. (7.3), show that $I = A_{L1}^2 + A_{L2}^2 + 2A_{L1}A_{L2}\cos(\phi_1 - \phi_2)$, where I is the light intensity of L.

7.2. Verify Eq. (7.8).

7.3. Using Eqs. (7.9) and (7.10), show that the difference in phase between the object and reference beams is $2n\pi$ ($n = 0,1,2, ...$) for the maximum light intensity and $(2n + 1)\pi$ ($n = 0,1,2, ...$) for the minimum light intensity. Then verify Eq. (7.11).

7.4. Verify Eq. (7.18).

7.5. Verify Eq. (7.19).

7.6. Verify Eq. (7.20).

7.7. Determine the deflection curve along a radial line for a clamped circular plate subjected to a lateral loading. The double-exposure holographic fringe pattern (the directions of illumination and observation are normal to the plate) is shown in Fig. 7.7. The radius (measured from the center of bolts to the plate center) of the plate is 3.5 in., and the wavelength is 24.91 μin.

REFERENCES

[1] R. Jones and C. Wykes, *Holographic and Speckle Interferometry: A Discussion of the Theory, Practice and Application of the Techniques,* Cambridge University Press, New York, 1983.

[2] W. F. Ranson, M. A. Sutton, and W. H. Peters, Holographic and laser speckle interferometry, Chapter 8 in *Handbook on Experimental Mechanics,* A. S. Kobayashi, ed., Prentice Hall, Upper Saddle River, N.J., 1987.

[3] C. E. Taylor, Holography, Chapter 7 in *Manual of Engineering Stress Analysis,* A. S. Kobayashi, ed., Prentice Hall, Upper Saddle River, N.J., 1982.

[4] R. E. Rowlands and I. M. Daniel, Application of holography to anisotropic composite plates, *Exp. mech.,* **12**(2), 75–82, 1972.

[5] R. C. Sampson, Holographic-interferometry applications in experimental mechanics, *Exp. Mech.* **10**(8), 313–320, 1970.

[6] R. L. Powell and K. A. Stetson, Interferometric vibration analysis by wave front reconstruction, *J. Opt. Soc. Amer.,* **55**, 1593–1598, 1965.

[7] C. R. Hazell and S. D. Liem, Vibration analysis of plates by real-time stroboscopic holography, *Exp. Mech.,* **13**, 339–344, 1973.

[8] D. B. Neumann, Comparative holography, *Hologr. Interferom. Speckle Metrol. Tech. Dig.* (OSA publ.), **MB2**, 1–4, 1980.

[9] P. K. Rastogi, Comparative holographic moiré interferometry, *Appl. Opt.,* **23**(6), 924–927, 1983.

[10] P. K. Rastogi, Comparative holographic interferometry: a nondestructive inspection system for detection of flaws, *Exp. Mech.,* **25**, 325–337, 1985.

C H A P T E R   8

Computer Data Acquisition and Control System

8.1   INTRODUCTION

To obtain the experimental results desired, one needs to collect data during the experiment and then process it in a desired format. The simplest method of acquiring experimental data might consist of one person reading one or several instruments and recording the observations on a data sheet. The subsequent data processing could be done by hand calculations, calculators, or sophisticated computer software. A technician logging temperatures of an oven on a piece of paper is an example of performing data acquisition. Another example is when one records the loads and corresponding displacements in a uniaxial tensile or compressive test. Later, the stress-strain could be established either by hand computations or by using computer software (e.g., Quattro Pro or Lotus-123), and Young's modulus can be determined for the material.

More sophisticated data acquisition systems should be used in rapid collection of a great bulk of data that must be processed later. Computers have become prevalent throughout our lives and have also become a major component in data acquisition and control systems. The purpose of this chapter is to introduce briefly the computer data acquisition and control system. With this system one can use the computer to gather, monitor, display in real time, analyze data, and to control the processes accurately for maximum efficiency. Since computer-based data acquisition and control systems differ greatly from each other and change rapidly, in this chapter we take a rather general approach. For more details, the reader should refer to Beckwith and Marangoni [2], Holman [3] and Yuan et al [4].

8.2   BASICS OF A COMPUTER DATA ACQUISITION AND CONTROL SYSTEM

Computer data acquisition and control system is the product of a combination of measurement technology, automatic control engineering, and computer technology. The system was developed in the early 1960s and was first used in chemical engineering to control production processes automatically. Since the 1970s, with the development of electronic technology, computer data acquisition and control systems have been widely used not only in laboratory and industry applications, but also in economics and even in our daily lives. In industry applications, the system is widely used to control, monitor, and manage production processes. Microcomputers and personal computers have become major components in computerized data acquisition and control systems, due to their high reliability and high price-to-performance ratio. Depending on the skill level of the user and the functions of the products, the computer may take on different roles. Computer data acquisition and control systems can be classified as those that plug directly into a computer and those that stand alone and interface to a computer through a communication port. In the former, the data acquisition cards or boards are plugged directly into the computer bus within the computer case. Two major advantages of plug-in data acquisition and control systems are speed (the cards are connected directly to the computer bus) and low cost (packaging and power overhead are eliminated). Stand-alone or communication-based data acquisition and control systems can range from data loggers to remote intelligent control systems. They are well suited for high-channel-count applications and are not restricted by the size of the computer, since the system is not part of the computer. Stand-alone systems can often be arranged in a mul-tidrop fashion, with many units connected together as a network. In either system, computer software is required to tell the computer how to handle the data collected.

Overall, computer data acquisition and control systems have the advantages of low cost, compact size, multifunctions, and large measurement range. Some typical functions are:

1. *Automatic operation.* The system can operate automatically according to the preset program once the operator gives the corresponding command.

2. *Automatic selection.* The system can select automatically the object to be measured, the measurement range, and the gain and frequency range necessary to achieve the optimum condition, so that the highest precision can be achieved during the measurements.

3. *Result judgment.* The system can make judgments as to whether or not the results measured are correct according to preset standards.

4. *Automatic adjustment.* The system can be adjusted to zero automatically. It can also self-adjust, according to preset standards, to eliminate some external inter-ference. Self-adjustment reduces measurement errors.

5. *Data processing.* The system can perform mathematic operations, error analysis, and unit conversion.

6. *Automatic control.* The system can automatically control the data acquisition process, make decisions on the control strategy, and control the active elements to perform the required operations.

7. *Trouble alarming.* The system can identify problems that occur during the entire data acquisition and control process.

8. *Network communication.* The system can use standard interfaces to communi-cate with various devices located externally to the computer or with other computer data acquisition and control systems.

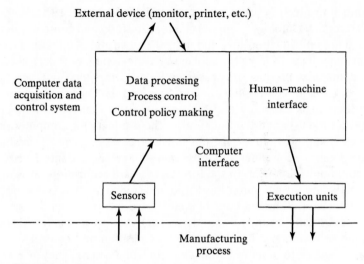

FIGURE 8.1   Computer data acquisition and control system in a manufacturing process.

Figure 8.1 shows a general computer data acquisition and control system in common manufacturing processes. In principle, the system can be divided into three major parts:

1. *Data acquisition:* to collect the data from the instrument transducers, then feed them to the computer (at the same time, the computer controls the data acquisition process).

2. *Data processing:* to process measured signals so that data which correspond to set parameters may be obtained, or decisions and judgments made.

3. *Data output:* to display the results on the monitor in graphic form or to print out results, and simultaneously, to send a control signal to the control mechanism units according to the decision made during the data processing stage.

8.3   COMPONENTS OF A COMPUTER DATA ACQUISITION AND CONTROL SYSTEM

Generally speaking, computer data acquisition and control systems consist of hardware and software. The hardware provides the ability to interface with a wide variety of analog signals, such as electric currents, voltages, and thermal couples. The software includes such features as graphic and tabular display of real-time data, control of alarm conditions, data logging to printer or disk, and control of processes. Figure 8.2 shows the hardware of a typical computer data acquisition and control system. According to their function in the system, the major elements of any computer data acquisition and control system are computer, input and output (I/O) channels, general peripheral devices located externally to the computer, interface circuit, dashboard, and system bus.

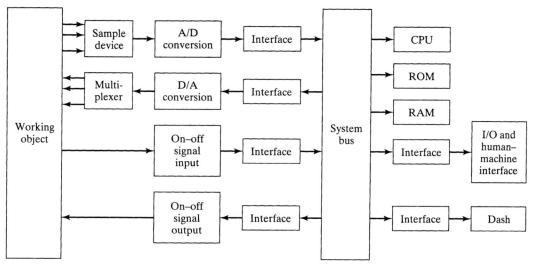

FIGURE 8.2   Hardware components in a computer data acquisition and control system.

Computer.   The computer has a CPU (central processing unit), ROM (read-only memory), and RAM (random access memory). The CPU, the heart of the computer, controls all aspects of the computing process. For example, an 80386 CPU typically operates at up to 33 MHz and offers a 32-bit bus. After the computer collects a variety of parameters and data have been input, the computer automatically processes and analyzes data and performs computations using computer software. If the purpose is to measure these parameters, the computer will give accurate results corresponding to the parameters being measured. If the objective of measuring the parameters is to control certain production process, the computer will make relevant control decisions or adjustments, then give control commands via output channels. Several types of computers are available. The type to use in a computer data acquisition and control system depends largely on the complexity of the tasks to be performed and on the technical requirements. Currently, most computer data acquisition and control systems use a personal computer as the host computer; a few systems use a workstation as the host computer.

Input and Output (I/O) Channels.   Input and output channels are channels through which the computer communicates with external devices or systems in the exchange of information (including data transmission and conversion). An input or output channel can be classified as an analog signal input channel, an analog signal output channel, an on–off signal input channel, or an on–off output channel. Analog signal input channels with analog-to-digital (A/D) converters can be used to connect transducers whose output signals are analog or as the input ends for analog voltage or current signals. An analog signal output channel has a digital-to-analog (D/A) converter, which makes it possible for the computer to control mechanical equipment or output devices operating in analog form. An on–off signal input channel is used to receive external information in on–off form, such as the on–off signal produced by a relay. In some real-time monitored control systems, an on–off signal is input to represent the state of overflow, alarm, and pole conversion, and to inform the computer to respond accordingly. An on–off signal can also be input into a computer in terms of codes. The code could be a command to ask the computer to perform an operation or could simply be digital data. An on–off signal output channel is often used to control on–off operating devices such as relays and steppers. It can also be used to output information in code form.

External or Peripheral Equipment. According to its function, peripheral equipment can be classified as input devices, output devices, or external storage. Input devices are used primarily in input programs and data. The keyboard is a common input device. Output devices are used primarily to provide an operator with various information and data in easily accepted forms such as in numbers, curves, and characters. Commonly used output devices include printers, monitors, and recorders. External storage equipment such as magnetic tape or magnetic disk recorder, is used primarily to store system programs and related data. The required numbers of external storage devices depend on the performance requirement of a system and affect system cost. If a personal computer is the host computer of a computer data acquisition and control system, and if the data size is not very large, one can use the internal storage of the computer to store the data, reducing system cost.

Interface Circuit. Input and output channels and external devices are connected to a computer via an interface. Thus the interface circuit plays the role of a vehicle that carries out information interchange among the computer, input and output channels, and external devices. Interface circuit design is a key step in a data acquisition and control system whose host computer is a personal computer. Typically, two types of interface systems, plug-in and stand-alone, are available. A plug-in interface system plugs into a computer directly and is designed for a specific computer type, such as an IBM PC or IBM-compatible computer. A stand-alone interface system, independent of computer type, can talk to any computer with a standard serial communication port.

Dash. A dash can be considered a special external device to realize human–machine interface. Through the dash, an operator can monitor the operating status of the system and modify parameters to adjust operation of the system if it is necessary. At a minimum, the dash consists of an input device such as a keyboard and a display device such as a CRT (cathode-ray tube). Some computer data acquisition and control systems use the computer keyboard and computer monitor as the dash.

System Bus. As mentioned earlier, an interface circuit plays the role of a vehicle that carries out information interchange among the computer, input and output channels, and various external devices. The system bus is the foundation that connects all components together, as shown in Fig. 8.3. Choosing the system bus is very important because it will affect not only system performance, but also system cost and the time required to build a complete system. For example, if you develop a computer data acquisition and control system starting with the CPU, you may use a system bus

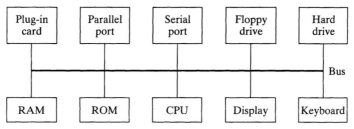

FIGURE 8.3 Components on a personal computer bus.

designed for a special purpose or a dedicated computer programmed for a specific application. Both are simple and inexpensive. On the other hand, you can use a computer bus as the system bus if you already have a personal computer. You can either plug your own designed hardware units into the computer or purchase the hardware units available in the market. In this way, you can build a computer data acquisition and control system quickly. Standard system buses such as the STD, S-100, VME, and DATA BUS are available.

Software can be classified as either application software or system software. *Application software* is developed by the system designer according to the requirements for measurement and control. Developing application software accounts for most of the time spent in software development. For control systems, application software is a direct control program, and all other system software is under its direct influence. Therefore, the quality of application software will greatly affect the accuracy and efficiency of a system. In general, *system software* consists of an operation system, monitor program, program design language, compiler program, and debug program. If you buy a personal computer as the component of the system, most of the necessary system software is available. You can modify the existing system software according to your applications. In some cases you need to develop your own system software.

Software is the bridge for connecting an operator's thoughts with the hardware. Although one cannot run software without hardware, the quality of software has a large impact on normal operation of a computer data acquisition system and on full utilization of the hardware functions. In the development of a computer data acquisition and control system, the time spent on software development may correspond to the time spent on hardware development.

8.4 ANALOG SIGNAL INPUT AND OUTPUT CHANNELS

In data acquisition and control processes, most original signals, such as temperature, pressure, flow, sound, and force, are in analog form. An *analog signal* is one that varies with time continuously; however, the computer can process data only in digital form. Therefore, computer-based data acquisition and control systems should have the ability to sample the analog signal and digitize the data (A/D conversion), then perform data processing and/or to convert the results in analog form in order to control external reaction mechanisms such as steppers. A block diagram for a general computer data acquisition and control system is shown in Fig. 8.4. The first task mentioned above

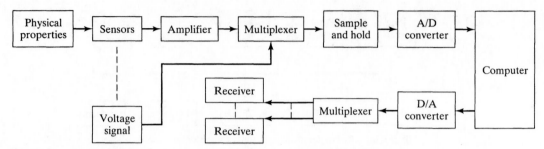

FIGURE 8.4   Computer-based analog signal input and output system block diagram.

is accomplished via an analog signal channel; the last task is carried out by the analog output channel. Some basic concepts are introduced in this section.

Sampling and Digitizing

The analog signal is a continuous function of time [i.e., $f(t)$], as shown in Fig. 8.5a. It is necessary to sample the continuous function in order to feed it into a computer to process it. Assume that the sampling period is T, which is defined as the *time difference* between the $(N+1)$th and Nth samplings, namely,

$$T = t(N + 1) - t(N) \tag{8.1}$$

The sampling frequency f is defined as

$$f = \frac{1}{T} \tag{8.2}$$

Sampling a continuous function $f(t)$ means to get the values of $f(t)$ periodically; thus $f(t)$ is represented by discrete data $f(nT)$, as shown in Fig. 8. 5b. To represent the analog signal by the sampled data without losing any information in the original signal, the sampling frequency must be greater than or at least equal to twice the highest frequency of the analog signal. This is called the *sampling theorem* or *Nyquist theorem*.

Sampled data are discrete in time but still are values of the analog signal. *Digitizing* converts these discrete analog signal into numbers that the computer can accept and understand. The high–low, yes–no, and on–off sequencing of digital devices suggests the use of a binary system for counting. For simplicity and illustration, assume that the computer can accept 4-bit words; it can only accept and understand digital codes that range from 0000 to 1111, a total of 16 codes. Every digital code corresponds to a definite analog level. Sampled signals often are not exactly equal to the analog level represented by a certain code, as shown in Fig. 8.5c. Thus the signal sampled is represented by the digital code nearest the point shown in Fig. 8.5c. For example, digital code 0011 represents sampled data points 5, 6, 7, and 10, although the actual analog signal values are not equal to each other for these points. Thus errors are introduced in the digitizing process. This procedure, digitizing the discrete sampled analog signal, is completed by an A/D converter (discussed later). In Fig. 8. 5c, symbols represent the sampled data points, and the corresponding vertical lines represent the digitized values. These digitized sampling signals can be fed into a computer for processing later.

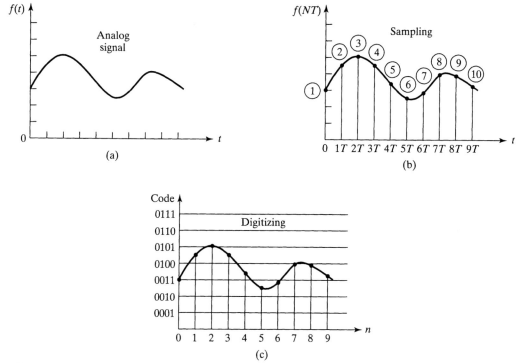

FIGURE 8.5 Sampling and digitizing an analog signal.

Obviously, the accuracy of digitizing or the resolution of the data acquisition will be increased greatly if an 8- or 16-bit word converter is used in the computer data acquisition and control system. However, the cost of the system will increase as well.

Simple Low-Pass Filter

It should be pointed out that unwanted analog signals, also called *noise,* are picked up during the sampling process, especially in industrial environments. This results in distortion and masking of the true signal. It is often possible, through the use of appropriate circuity or an appropriate digital filter program, to filter out some or all of the unwanted noise. Figure 8. 6a shows a simple resistance–capacitor (*RC*) low-pass filter. The response of a more complicated low-pass filter is schematically shown in Fig. 8. 6b. The filter will attenuate signal frequency above the cutoff frequency computed by the values of the resistance and the capacitance if an *RC* low-pass filter is used, or input by the user if a digital low-pass filter is used.

Analog Signal Input

The main task of an analog signal input channel is to convert analog signals detected from the measurements by various transducers or sensors to digital signals, and then to feed them into a computer via an interface. In general, an analog signal channel con-

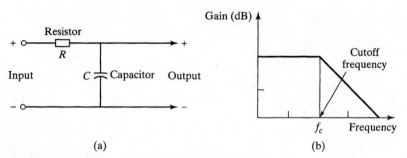

FIGURE 8.6   (a) Simple low-pass filter; (b) gain–frequency diagram for a low-pass filter.

sists of signal-processing circuits, multiplexers/switchers, amplifiers, and analog-to-digital (A/D) converters. It also contains interface circuits and control circuits, to provide a connection with a computer. A typical block diagram of an analog signal input channel is shown in Fig. 8.7.

The function of a signal processing circuit may vary with the transducers or sensors used to detect or to sense the measured quantity. Signal processing in general includes small-signal amplification, signal damping, signal filtering, impedance match, level conversion, nonlinear compensation, and electric current-to-voltage conversion, and other functions, some of which are explained in more detail later in the chapter.

Multiplexer.   Generally speaking, only one analog-to-digital converter is used in a general multianalog signal data acquisition system. It is necessary to use the multiplexer in an on–off setting, following a certain sequence, or in a random setting, to allow analog signals in different input channels to pass through one at a time. The multiplexer, also called a multiple switcher, can be placed in front of a common amplifier or analog-to-digital converter. An ideal (nonexisting) multiplexer has an infinite open-circuit resistance, a zero closed-circuit resistance, quick on–off switch rate, low noise, long life, and high reliability. The multiplexer is classified as either a mechanical contact switch or an electronic switch. Examples of mechanical contact switches are the dry reed relay and the vibrator relay. Examples of electronic switches are the transistor

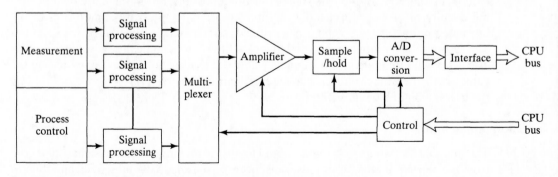

FIGURE 8.7   General analog signal input channel.

and the field-effect transistor. A dry reed relay is an ideal mechanical contact switch; it has a low closed-circuit resistance ($< 0.1\ \Omega$), a high open circuit resistance (up to 10 G Ω), and a long working life (up to 10^6 to 10^7 times). Occasionally, however, the contact point may not be separated completely. This kind of relay is suitable for applications in sampling units for small-amplitude signals with medium speed. (The conversion rate is 10 to 400 samples/second.) On the other hand, an electronic switcher has a very high on–off frequency (the conversion rate can be up to 1000 samples/second), small size, and a long life, but it cannot be used to measure a small-amplitude signal accurately because its open-circuit resistance is relatively high (between 5 Ω and 1 kΩ), and the driving part is independent of the switch elements. Today, it is common to use integrated multiple electronic switches, except for cases in which small-amplitude signals are to be converted accurately. In such cases, a low- or medium-speed high-precision data acquisition unit should be used.

Sample-and-Hold Circuit. When the frequency of a measured analog signal is high, it is necessary for the input channel to react in time. It takes time, however, to complete the analog-to-digital conversion process. Therefore, the numerical data converted cannot represent the true analog signal level at the instant the command is issued if the input analog signal changes appreciably during the A/D conversion process. To overcome this shortcoming, a sample-and-hold (S/H) circuit should be used, generally placed in front of the A/D converter. An S/H circuit can sample the analog signal quickly enough to capture the appreciated change of the signal, then hold the sampled signal during an analog-to-digital conversion process. In this way, the allowed frequency of an input signal is raised. Figure 8.8 shows a simple sample-and-hold circuit, which consists of an analog switch (K), a storage capacitor (C), and a buffer amplifier. When a control signal turns the switch on, data sampling begins. The input signal is passed through the resistance R and charges the capacitor C. The output signal of an S/H circuit varies with the change in input signal. In general, the shorter the charge time, the better the S/H circuit. When a control signal turns the switch off, holding begins. The input voltage at the precise time the switch is turned off is held in the capacitor.

An S/H circuit is a complicated circuit that has several function parameters, two important ones being capture time and voltage decrease rate. *Capture time,* defined as the minimum time necessary for an input signal to reach a value within a prescribed percentage error, is a function of the charge time constant of the capacitor and the

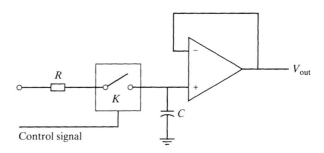

FIGURE 8.8 Simple sample-and-hold circuit.

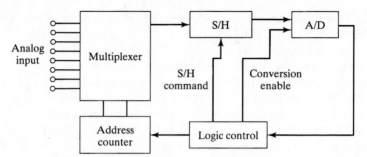

FIGURE 8.9   Multichannel computer data acquisition system.

swing rate and setup time of the amplifier. *Voltage decrease rate* is defined as the discharge rate of the storage capacitor during the holding period, a function of the capacitance of the capacitor *C*, leakage resistance, the input resistance of the buffer amplifier, and the switch-off resistance of the analog switch *K*. A multichannel computer data acquisition system with multiple converters and S/H circuit is shown in Fig. 8.9.

Amplifier.   To achieve the accuracy required for A/D conversion, an A/D converter must amplify the input signal picked up by various transducers or sensors to the required input electric level. The range of the electric level is 0 to 10 V or 0 to 5 V for a unipolar unit and −5 to +5 V or −10 to +10 V for a bipolar unit. Three basic amplifiers are commonly used: an inverting amplifier, a noninverting or in-phase amplifier, and a difference or differential amplifier. Each is shown schematically in Fig. 8. 10.

For an in-phase amplifier, the analog signal is input from its in-phase terminal, and the output signal is in phase with the input signal. Similarly, the analog signal is input from the inverting end and the output signal is out of phase with the input signal for a inverting amplifier. For a differential amplifier, signals are input from both the in-phase and out-of-phase ends, and the output signal is proportional to the difference between the two input signals. This type of amplifier can be used to perform a subtraction operation. When many transducers are involved in the application, a common amplifier may be used for multiple channels. If different gains are required for differ-

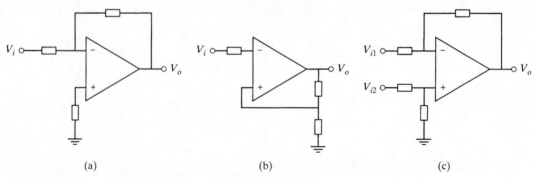

FIGURE 8.10   (a) Inverting amplifier; (b) in-phase amplifier; (c) difference amplifier.

ent channels, one can design a programmable amplifier. In such a case, the computer is used to control the amplifier's close-loop gain.

Integrated A/D Chip.   The integrated analog-to-digital conversion chip has the following functional pins: data output, conversion enable (input), and end of conversion (output). The interface used to connect an A/D chip with a CPU deals with the connection between the three types of pins and the CPU. The A/D chip typically has a chip enable pin, whose connection format is the same as that of all other I/O interface chips. For an A/D chip with multiplexer, it is necessary to connect the channel conversion control signal correctly. When utilizing A/D chips, the following three aspects should be considered:

1. The A/D chip compatiblity with the system bus of the microcomputer or PC: whether the data output lines of the A/D chip can be connected directly to the CPU data bus.

2. The trigger A/D conversion mode: requires an external trigger conversion signal, generally determined by the CPU. Different A/D chips need different types of trigger conversion signals. There are two types of trigger signals in practice: the pulse trigger signal and the electric level control signal.

3. The conversion end signal mode: indicated when the A/D chip outputs an index electric level indicating the end of the conversion process to inform the CPU to read the data converted. The CPU reads the data using either interrupt or inquiry mode. When interrupt mode is used, the conversion end signal is sent to the interrupt request pin of the CPU or an I/O interface, with the function of permit interrupt, to ask the CPU to interrupt. In response to the interrupt request signal, the CPU uses the A/D converter to read the data via the corresponding interrupt service routine. If inquiry mode is used, the conversion end signal is sent to a certain bit of the data bus or of a parallel I/O interface. The CPU requires the bit to judge continuously whether the conversion process has been finished. If the process has been finished, the CPU reads the results. Which mode is to be used in a computer data acquisition and control system depends largely on the A/D converting rate required and the user's software. Inquiry mode takes CPU time but is simple in operation. Thus inquiry mode is often used in data acquisition with high A/D converting rates. On the other hand, interrupt mode is more suitable for applications using A/D converters with low or medium rates of A/D conversion.

Analog Signal Output

The analog signal output channel converts the digital outputs by the computer into analog signals. The key part of an analog signal output channel is the D/A (digital-to-analog) converter. When performing digital-to-analog conversions, the digital codes to be converted are put into the data input end of the D/A converter, and the corresponding electric currents or voltages are obtained at the output end of the D/A converter. However, the data inputs of a D/A converter may not be connected directly to the data bus of a microcomputer or PC for certain type of D/A converters, since in practical applications, the converted analog signal needs to hold for awhile in order to detect or control the measurand conveniently. Therefore, the input end should be connected with the CPU data bus via a latch circuit for the D/A chip without an input data register.

Digital-to-Analog Converter

The inputs of a digital-to-analog converter are digital signals; the outputs are analog electric current or voltage signals proportional to the corresponding digital numbers. A D/A converter has better anti-interference ability than that of an A/D converter because its inputs, the digital numbers, do not drift with time and temperature. Two commonly used D/A conversion modes are introduced in this section: binary-weighted resistance mode and ladder network mode.

Binary-Weighted Resistance D/A Converter. A binary-weighted resistance mode with eight switches is shown schematically in Fig. 8.11. This converter consists of four parts: (1) eight switches corresponding to an 8-bit word; (2) reference voltage V_{ref}; (3) weighted resistance network; and (4) summation unit, such as a summing amplifier or op-amplifier. The 8-bit register stores a binary number N, which is to be converted. Thus

$$N = a_7 2^7 + a_6 2^6 + \cdots + a_1 2^1 + a_0 2^0 = \sum_{i=0}^{7} a_i 2^i \tag{8.3}$$

where each a_i (take the value of either 0 or 1) controls a switch S_i. When $a_i = 1$, S_i is connected to V_{ref}; when $a_i = 0$, S_i is connected to ground.

As is shown in Fig. 8.11, switch 7, connected to a resistor with resistance R, corresponds to the most significant bit (MSB), a_7, and switch 0, connected to a resistor with resistance $2^8 R$, corresponds to the least significant bit (LSB), a_0. The precision and stability of the resistance for the least significant bit are less important than those for the higher or most significant bits. By the superposition principle of the circuit, the electric current I to be sent to the summing amplifier is

$$I = a_7 \frac{V_{ref}}{R} + a_6 \frac{V_{ref}}{2R} + \cdots + a_1 \frac{V_{ref}}{2^6 R} + a_0 \frac{V_{ref}}{2^7 R}$$

$$= \frac{V_{ref}}{2^7 R} \sum_{i=0}^{7} a_i 2^i = \frac{V_{ref} N}{2^7 R} \tag{8.4}$$

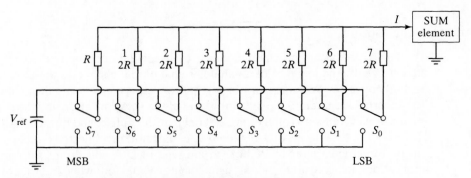

FIGURE 8.11 Eight-bit binary-weighted resistance D/A converter.

As can be seen, the output current is proportional to the input digital number N. The disadvantage of a binary-weighted resistance converter is its limited integrity, due to its wide range of resistance variations.

R–$2R$ Ladder Resistance Network Type.   An R–$2R$ ladder network D/A converter, on the other hand, has only two types of resistor: R or $2R$. It is simple and can be integrated easily and thus is widely used in a single D/A converter chip (e.g., in DAC 0832, an 8-bit D/A converter). An R–$2R$ ladder network D/A converter consists of four parts, similar to those of a binary-weighted resistance D/A converter. The only difference between the two types of D/A converter is that the weighted resistance network of the binary-weighted type is replaced in the second type by an R–$2R$ ladder network, as shown in Fig. 8.12. It can be seen that the resistance connecting each switch is $2R$.

Assume that if $a_i = 1$, switch S_i is connected to V_{ref}; and that if $a_i = 0$, switch S_i is connected to ground. By the linear superposition principle, the electric current I to be input to the op-amplifier is

$$
\begin{aligned}
I &= a_7 I_7 + a_6 I_6 + \cdots + a_1 I_1 + a_0 I_0 \\
&= \frac{a_7 I_f}{2^1} + \frac{a_6 I_f}{2^2} + \cdots + \frac{a_1 I_f}{2^7} + \frac{a_0 I_f}{2^8} \\
&= \frac{1}{2^8} I_f (a_7 2^7 + a_6 2^6 + \cdots + a_1 2^1 + a_0 2^0) \\
&= \frac{1}{2^8} \left(\frac{V_{\mathrm{ref}}}{3R} \right) \sum_{i=0}^{7} a_i 2^i = 2^{-8} \left(\frac{V_{\mathrm{ref}}}{3R} \right) N
\end{aligned}
\tag{8.5}
$$

where $I_f = V_{\mathrm{ref}}/3R$. Again, the output current from the D/A converter is proportional to the digital number N.

In practice, the desired output from the D/A converter is voltage instead of current. This can be achieved easily by connecting another op-amplifier, shown in Fig. 8.13. In Fig. 8.13a, the output voltage V_{out} from the op-amplifier is out of phase with the output current of the D/A converter; that is,

$$
V_{\mathrm{out}} = -IR
\tag{8.6}
$$

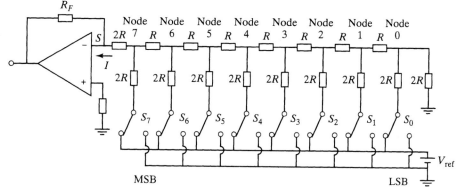

FIGURE 8.12   Ladder resistance network D/A converter.

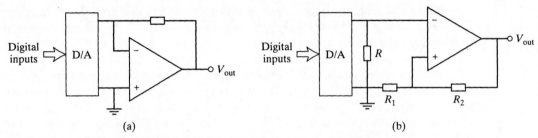

FIGURE 8.13 Voltage output of a D/A converter: (a) inverting output; (b) in-phase output.

whereas in Fig. 8.13b, the output voltage is in phase with the electric current from the D/A converter:

$$V_{out} = IR\left(1 + \frac{R_2}{R_1}\right) \tag{8.7}$$

Analog-to-Digital Converter

An A/D converter converts an analog signal to n-bit binary numbers, or a continuous signal to 2^n different digital numbers. An A/D converter may be created by adding a voltage or analog comparator to a D/A converter and modifying the corresponding operating routines.

Voltage Comparator Circuit. Figure 8.14 is a schematic of a voltage comparator circuit that has the following features: (1) A small voltage difference between V_{in} and V_{ref} swings the output to a limit permitted by the power supply, where V_{in} is the input voltage and V_{ref} is the reference voltage being set to a desired reference values; (2) the impedances at the positive and negative ends are approximately equal if $R_1 \approx R_2 R_3 / (R_2 + R_3)$; (3) diodes are used to limit the differential input; and (4) the output voltage is in phase with the input voltage when $V_{in} > V_{out}$, and the output voltage is out of phase with the input voltage when $V_{in} < V_{out}$. This feature of the comparator provides an output indication for the relation between the input voltage and the reference voltage. The analog comparator may drift with time and temperature since the inputs are analog signals, not digital numbers.

There are several A/D converting methods; only two of the commonly used methods are introduced here: the counter method and the gradually approaching method.

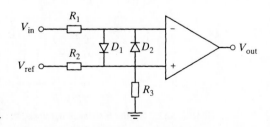

FIGURE 8.14 Voltage comparator.

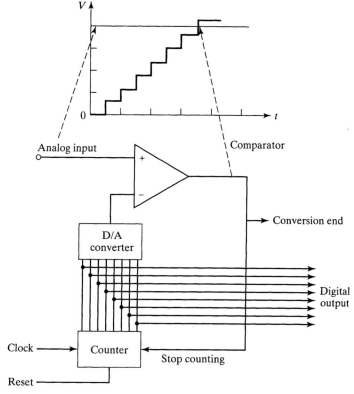

FIGURE 8.15   Counter type of A/D conversion.

Counter Method.   The simplest and cheapest A/D converter is a counter type of converter, shown in Fig. 8.15. One counter controls one A/D converter. First the counter is set to zero, then the D/A converter outputs a rising ladder voltage. The input analog signal and the rising ladder voltage signal are sent to the analog comparator for comparison. An indication signal to stop the counting is produced when the two signals are equal or when the difference between the two signals is small enough within a permitted error range. The output of the D/A converter at this moment is the approximate value of the sampled analog signal, and the corresponding digital value is given by the counter.

This process can be seen to be similar to the process of weighing an object by increasing the smallest weights of the balance one by one (only one type of weight). The conversion time for a counter A/D converter is relatively long and may vary greatly. When the analog input is low, a small number of steps are required to reach equality; when the analog input is high, a large number of steps are required.

Gradually Approaching Method.   Figure 8.16 describes an A/D converter using the gradually approaching method. The A/D converter also consists of a D/A converter but is controlled by a register instead of a counter. In counter A/D conversion,

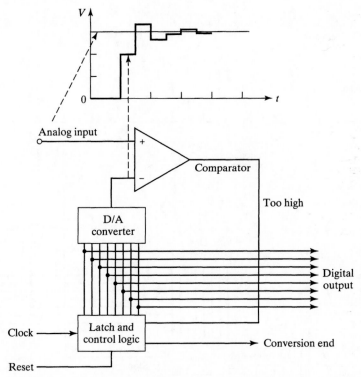

FIGURE 8.16   Gradually approaching type of A/D conversion.

comparison begins at the lowest bit in sequentially increasing manner. Converse to the counter method, the gradually approaching method starts at the highest bit and moves in sequentially decreasing order. At the beginning, the register is set to zero. During the conversion process, the highest bit (MSB) is set to 1. The analog input is compared with the output of the D/A converter by the comparator; 1 is kept if the output of the D/A converter is less than the input analog signal, otherwise, 1 is replaced by 0 for the bit. The process then goes to the next-highest bit and repeats the process until the lowest bit (LSB) is reached. The conversion process is then finished for these sampled data. The digital number in the register is the corresponding approximate value of the signal being sampled.

   This process can also be viewed in terms of the process used to weigh an object. First the heaviest weight is placed on a balance and is taken away if too heavy and replaced by a lighter weight; if lighter, the weight is kept on the balance and the next-heaviest weight is added. The process is repeated until all weights have been tried.

   Both the the counter and gradually approaching methods are of the feedback type. For a 8-bit A/D converter, the gradually approaching method needs only eight comparisons to finish the conversion, but times for comparisons using the counter method are variable. The maximum number of attempts needed for the counter method may reach as high as 256. The gradually approaching method is used in medium A/D conversion rates of 8- or 16-bit A/D converters.

8.5 INTERFACE FOR A COMPUTER DATA ACQUISITION AND CONTROL SYSTEM

As mentioned earlier, in a computer data acquisition and control system, the heart of the computer is the central processing unit (CPU). The CPU serves as the control center for interchange of information with internal storage, and for frequent interchange of information with various peripheral equipment via input and output channels. The CPU accepts input data in digital form, through either command instructions or numerical data. It then routes the inputs to programmed destinations over buses to display them in graphics or numbers, to memorize inputs, to control devices, and so on. For certain media or large computer data acquisition and control systems, two or more computers may be involved, so information interchanges between computers or between systems are often necessary. The information interchanges cited above are called *communications*. The CPU communicates with peripheral equipment by utilizing input programs (software) and interface circuits (hardware). The following four types of communications are frequently used in practice: inquiry mode, immediate mode, interrupt mode, and direct memory access mode. The process of connecting an I/O device directly to memory is known as *direct memory access* (DMA). The hardware that controls DMA is the *DMA controller*. An example is the 8237 chip on the system board of an IBM PC.

1. *Inquiry mode.* The CPU interchanges information with peripheral equipment by running a piece of program. The CPU constantly makes inquiries as to the working status of the peripheral device to see whether information interchange is allowed. If the answer is "yes," information interchange takes place with the device once per inquiry.
2. *Immediate mode.* Immediate mode is also called *synchronous mode*. The information interchange is made immediately after receiving the information. It is necessary to accurately compute the time used for the information interchange between the CPU and peripheral device if immediate mode is to be used in interchanging the information. Synchronous mode is used for information interchange only in cases when it can be ensured that losing information or interchanging the same information twice or more will never happen.
3. *Interrupt mode.* In interrupt mode, peripheral equipment asks the CPU to interrupt, and information interchange is made each time the CPU responds to the request.
4. *DMA mode.* In DMA mode, peripheral devices such as the disk driver interchange information directly with internal storage.

To adopt any one of the first three modes, the CPU needs to run a program and then interchange information only when it is allowed to do so. Since the time for interchange information is only a portion of the time needed for execution of the program, this will affect the information interchange speed between the CPU and peripheral equipment. While in DMA mode, the peripheral device sends a DMA request signal to the CPU first; after receiving the request and in response, the CPU then frees the system bus and lets the DMA controller accept and control the system bus. Thus information is interchanged directly between the peripheral device and internal storage via the

DMA controller. Once this process is finished, the DMA controller frees the system bus and informs the CPU immediately.

The CPU communicates with various peripheral devices in either parallel or serial format. The most common parallel data acquisition bus is IEEE-488, and the standard serial buses are RS-232, RS-422, and RS-485. *Parallel format* is defined as several bits of a data being transmitted at the same time. On the other hand, *serial format* means to send several bits of the data one by one in the same wire according to a specified sequence. The transmission rate is much higher in parallel than in serial if the response speed for both devices is the same. In long-distance transmission, however, fewer signal wires are needed for serial transmission than for its counterpart, so serial transmission is frequently used in long-distance applications to reduce the cost of the system. An interface is necessary for both transmission formats. In modern computer data acquisition and control systems, the analog signals being detected from the measured quantities or the processes are, in general, changing continuously. To monitor these processes, it is often necessary to do high-speed measurements for various parameters integratively. Standard interfaces are recommended to connect various measurement instruments, computer, peripheral devices, and sensors to form a truly integrated system that is compatible with the work speed and control form. Generally speaking, the interface has the following functions:

1. *Input:* converts the data or information transmitted to the system from sensors, instruments, and peripheral equipment into a format compatible with the system.
2. *Output:* converts the data or information to be sent to the peripheral devices by the system into a format compatible with the corresponding peripheral equipment.
3. *Control pulses:* produces appropriate input, output, data transmission in synchronous or asynchronous pulses, clock pulse, and control-selecting on–off pulses, and allows the system to work compatibly.
4. *Interrupt process:* detects and processes the interrupt request signal from the peripheral devices.

8.6   SUMMARY REMARKS

Today, a variety of integrated cards and interface boards with supporting software are available in the market, which makes developing a computer data acquisition and control system much easier. A few such cards, for example, are the high-performance analog-to-digital interface board (model DAS-20) for IBM PC/XT/AT and compatible systems which can be operated in 16-channel single-ended or 8-channel differential input modes with an A/D conversion rate of 100,1000 samples per second. Another example is the PCL-718/818 board, a high-performance analog and digital interface board for IBM PC/XT/AT and compatible systems, ideal for industrial applications such as data acquisition, automatic testing and factory automation, and process control. Details may be found in reference 1.

In this chapter we considered the entire computer data acquisition and control system and explained some of its basic concepts. It is our hope that the material will serve as an introduction to readers interested in further study of the topic and will be useful in selecting among the integrated boards available in the marketplace.

PROBLEMS

8.1. What are the major elements of the hardware in a computer data acquisition and control system? What is the function of each element?

8.2. Explain briefly the working principle of a sample-and-hold circuit. Can an S/H circuit be placed in front of a multiplexer?

8.3. How many communication modes are there between a computer and peripheral devices in a typical computer data acquisition and control system? Which mode is fastest?

REFERENCES

[1] *The Data Acquisition Systems Handbook,* Vol. 28, Omega Engineering, 1992.

[2] T. G. Beckwith and R. D. Marangoni, *Mechanical Measurements,* 4th ed., Addison-Wesley, Reading, Mass., 1990, Chapters 7 and 8.

[3] J. P. Holman, *Experimental Methods for Engineers,* 5th ed., McGraw-Hill, New York, 1989, Chapter 14.

[4] Shenfang Yuan, Baoqi Tao, and Guo Liu, Development of high speed and large data flow data acquisition and control system and software design, *Comput. Autom. Meas. Control,* 4(1), 1994.

Index